NASA
Secrets,
The story of the Space Shuttle
Vehicles— Launching Satellites

ALBERT MONROE

authorHOUSE®

AuthorHouse™
1663 Liberty Drive
Bloomington, IN 47403
www.authorhouse.com
Phone: 833-262-8899

Published by AuthorHouse 07/11/2022

ISBN: 978-1-6655-6329-1 (sc)
ISBN: 978-1-6655-6328-4 (hc)
ISBN: 978-1-6655-6327-7 (e)

Library of Congress Control Number: 2022911891

# CONTENTS

# INTRODUCTION

Humans in the seventeenth, eighteenth, and nineteenth centuries could only look to the sun, moon, and stars and wonder what was out there—and how they could possibly get there. Once the telescope was invented, we could see further and determined there were seven planets out there, the Milky Way, and other astronomical bodies. Researchers and Scientists could measure the rotation of our planet and determine seasons, which had repeated their transits for millions and billions of years. Modern people can now look to the stars and navigate, which allows us to travel long distances across the globe with accuracy. In the twentieth and twenty-first centuries, with modern technology, we can not only see what's out there, but can even leave Earth and travel to outer space.

On our planet, we've traveled nearly everywhere. We've looked in every nook and cranny to study our natural wonders: the deserts, the mountains, caves, the poles. We've climbed to the highest mountain peaks and learned about everything we could. We've even tried to conquer the oceans to an extent. We've learned about every species we could on land and from the sea. We are still limited in our capacity to study the oceans, but with the advancement of modern technologies we've been able to go deeper and deeper to quench our thirst for knowledge. The only thing left is space, or as many calls it, the *final frontier*.

As technology continues to improve, humans can do things we've never done before. Our thirst to explore space keeps us interested, and many believe that one day, billions of years from now, it will not only be nice, but will be necessary for our survival. *We will need to find another home.* The so-called experts say nothing lasts forever, and in time our sun will burn out and become a dead star, and we'll not be able to support life on planet Earth.

Our species will die out unless we can find another suitable place to live. Scientists predict small animals and insects will be the only living things that may survive. None of us reading this book or anyone in the immediate future need to worry about this; however, there are possibilities that large meteorites could collide with Earth and destroy life, like what happened to the Dinosaurs. This is very unlikely, but as we have seen in movies, this doomsday scenario just might happen. If we are *reactive* rather than *initiative-taking*, we might just be too late.

This thirst to reach the final frontier only came about after German physicist Von Brun Brawner and Robert Goddard invented rockets, which made it possible to leave earth's atmosphere. After the Soviet Union put up Sputnik and John Glenn orbited the earth the space race focused on the moon, and a competition escalated between the US and the Soviet Union. Once President Kennedy gave us the green light to land a man on the moon and return him back home, the race became more than a pipe dream. There were many failures along the way, which cost numerous lives and billions of dollars, but once the Soviets saw the US would get to the surface of the moon before them—and it was costing them too much—they dropped out of the moon race.

What we needed to achieve our goals was an organization to run and monitor this large endeavor. The National Aeronautics and Space Agency (NASA) was created for this purpose. As a government entity with deep pockets, anything and everything was a go. Astronauts and their payloads were launched and carried into space, aboard our rockets, like the Atlas, Delta, and Titan vehicles. The Cold War was looming large, and our military was developing their own rockets, called the Intercontinental Cruise Ballistic Missiles, (ICBM) to deter the threat from the Soviets. As the technology improved, a bigger rocket was needed. The Saturn Five rocket—the biggest ever built—was the answer. After the moon landing in 1969 and the Cold War and expansion of the Soviet Union, things heated up until the collapse of the Berlin Wall between East and West Germany changed everything in 1989. The Soviet Union collapse, and a so-called Democratic government was formed in 1981.

NASA envisioned we could go further, but a crewed space vehicle with unlimited capabilities was needed. The space shuttle was invented and developed to serve this purpose, and six were built. Different names are used—some call them *orbiters*, some say *space vehicles* or *spacecraft*. They were manufactured in the Palmdale plant in Southern California, and the first to fly was the *Challenger* in 1981.

In this book I will discuss the shuttles and their purpose and capabilities. They were built to be reusable and are an engineering marvel. Building something like this had never been attempted before. NASA needed to expand and created several flight centers around the country with specific purposes and duties.

The book will cover NASA, its entities, and their sites. Satellites in orbit are essential for our safety, communication, security, and other purposes. Getting them there is not easy but keeping them there is the harder part. The unmanned rockets did this before the orbiters came online. Afterward, the shuttles did the heavy lifting and were still performing this duty until they Shuttle program were canceled in 2011. Some say, the program was costing the U.S. millions, and more each year to fund and run, and we were not getting our bang for our buck.

After leaving the Air Force, in 1985, I joined Lockheed Missiles and Space Company (now known as Lockheed-Martin) as a Satellited Operation trainer and worked on a joint Lockheed/NASA contract for ten years. We trained the Air Force officers along with Lockheed and NASA engineers, who launched satellites from the Blue Cube—a large blue building housed at Onizuka Air Force Station in Sunnyvale, California, formally Sunnyvale Air Station. My team and I also trained those engineers who were going through a shuttle orientation course on satellite operations, those launched from the Shuttle cargo bay once or orbit.

The book will cover every aspect of the shuttles. I'll be taking you through their life spans (I call it cradle to grave), how the shuttles are prepared for launch, and how they are launched. Then I will go into what gets the job done once the shuttle is in orbit. After the work is done, I will take you through how the shuttle leaves orbit and lands back here on earth. Last, I will talk about two major crashes and how astronauts go through training. The Hubble Space Telescope and the new James Webb Telescope and TDRSS will be covered as well.

Chapter 1 covers the *Challenger* tragedy that took place on January 28, 1986—an event many say could have been prevented. I mention a few secrets NASA did not want the public to know about. I talk about my participation in these endeavors and how, three thousand miles away in Sunnyvale, California, we were affected.

CHAPTER 1

The Daily Grind

I woke up early after tossing and turning all night. It was January 28, 1986—a beautiful Northern California day with blue skies, a wonderful day for a space shuttle launch. Three thousand miles away in Florida, the launch was scheduled out of Kennedy Space Center (KSC) in Cape Canaveral—our country's launch site. The orbiter, the *Challenger*, was going up after twelve months of delays, improvements, fixes, updates, and cost overruns.

On any other day, my main concern was beating the traffic, but this morning I started out uneasy. I couldn't put my finger on it, but something sinister danced in the back of my mind. Not expecting anything drastic to occur, I never could have imagined what was to unfold two hours later. Minor glitches are common. Shuttle systems and equipment breakdowns often occur during liftoff. But a major disaster was never envisioned.

I had recently retired from the air force after a lengthy career. Due to my background in training and development, I was assigned as a special mission's trainer at Lockheed Missile and Space Company. The department was founded to train personnel working at the Blue Cube, the five-story building where hundreds of our country's satellites were monitored and controlled. Commercial type Satellites was only being developed at that point.

It was about two blocks from our Training facility. With my

background and experience, I joined two major training teams. After six months of specialized and personalized training, we were certified and qualified to teach both the shuttle and training seminars.

Lockheed primarily made airplanes in their other plants when I was working there, but Sunnyvale had a large manufacturing division where satellites, like those for communications, were built. The Hubble Space Telescope, which is still in orbit, was built at the Sunnyvale plant.

Taking several years to complete construction, I was able to go down to the viewing area and see it being built. All the workers there wore the white bunny suits, and it was assembled in an exceptionally clean environment. To ensure cleanliness, all dust and dirt particles were removed in an ultra clen room environment.

Lockheed was just across the bay from Fremont, where I lived. Usually, I was not anxious to get going, but that morning was different. Little did I know, it was going to be a monumental day in history. Going to work was a daily hassle, especially crossing the San Mateo Bridge over to the west side of the bay. Depending on the time I left, there were holes in the traffic. Leaving at my regular time, traffic was steady. If I waited twenty more minutes, it could take close to an hour to get to work.

I was in the training department just over a year when NASA awarded Lockheed a large contract to provide course development and training classes. Over the next ten months, we trained hundreds of Blue Cube personnel, with a new class starting every two weeks.

I had thought about this day for months. Close to one hundred air force officers, engineers, and contractors had been prepared for the mission. Everyone was ready. I am sure those teachers and their students watching were glued to their TV sets due to the special astronaut on board. NASA had various large operational training facilities spread across the country.

Marshall Space Flight Center in Alabama was ready, and so was the Johnson Space Flight Center in Houston—the main command center. Of course, the launch team at KSC was full of anticipation. We had trained the large cadre of people supporting this mission, including the seven space shuttle astronauts on the flight. After months of downtime, NASA brass and the government considered this a significant mission.

But first, let me provide some background and details of the space program and our satellites before I digress back to that tragic day that killed seven astronauts and forever changed the space program.

As more satellites launched into space, problems arose. At the Blue Cube, they were launching satellites into both low and high orbits, depending on the mission. and it became evident that specialized training was the answer. Launching payloads from a rocket or from a shuttle's cargo bay was apples and oranges. Timelines had to be adhered to. The shuttle could be in complete daylight with temperatures way above freezing and plunged into total darkness five minutes later, at zero temperatures. Sunlight and darkness take no time with speeds more than seventeen thousand miles an hour.

For employment at the Blue Cube as satellite operators, one had to attend our six-week mission control training operations seminar. Passing it was crucial and mandatory to work there. Specialized subjects taught included orbital analysis, communications, command and control, Space Ground Link System, and many other functions. This course was intensive with only two chances to pass. Not everyone was suited for it, and two or three in every class failed.

For years unmanned rockets like the Delta, Atlas, and Titan launched our payloads into orbit. Due to weight restrictions, the satellites were small. They came in all shapes and packages and, depending on the mission, could be launched anywhere. Technology changed everything, especially in space. Communications became paramount.

Cell phones were evolving and became a critical need. Com Sat's, satellites for communication flooded space as they flew in a low Earth orbit. During the Cold War era, satellites and their special payloads provided a great asset to our military defense. Just imagine the capabilities we had over our adversaries. Big Brother is always watching!

Shuttle sitting on Launch Pad

Operations in Sunnyvale were performed by Lockheed engineers and others who worked under the NASA contract. For daily operations and control, five major positions needed to be trained: mission controllers, data controllers, orbit analysis, planner analysis, and ground controllers. Each position controlled certain aspects of the satellite or its function.

Hundreds of satellites were launched, such as meteorological/weather, communication, global positioning satellites (GPS), and Department of Defense (DOD) classified or surveillance. Circling the globe, the payloads are crucial for our safety on many fronts and save hundreds and millions of lives each year. We now can track storms which save lives. Two major storms come to mind.

Hurricanes Katrina and Ida can attest to this. The weather satellites monitor and track the climate, wind speed, pressure, and other aspects, such as weather patterns. Classified satellites were just that, and communication and GPS systems were more plentiful due to their function.

Payloads in a low Earth orbit up to one hundred fifty miles high had to be monitored constantly, while those in higher orbits could go for months without any contact. The air force officers oversaw each satellite or program, but the daily upkeep was done by the Lockheed mission controllers, who oversaw the other Lockheed contractors.

The data analysis group did daily data analysis. The orbit analysis team maintained the payload's orbit, and the planner analysts did plan. The ground controllers monitored the daily status and upkeep of the satellite, like battery charging.

After landing on the moon in 1969, we envisioned we could eventually work and live in space. Using rockets as a launch vehicle was very costly. We needed a new space platform that could take us to space and return to Earth and, more importantly, be reusable. These types of vehicles had been created and eventually, developed with so many manufactures taking part. In a great American manufacturing effort, numerous companies were involved, and six NASA orbiters were built, costing at that time billions of dollars. NASA brass, congressional folks from Florida, and some accountants agreed the long-term cost would be reduced after several flights.

Overruns were common, causing delays, but eventually the *Enterprise* was delivered to NASA, followed by the other five—*Columbia*, *Challenger*, *Explorer*, *Atlantis*, and *Discovery*. Officially named the Space Transportation System, the program started in 1981 with the first launch of the *Colombia* on April 12. The STS program lasted for thirty years until 2011 when it was canceled.

The *Enterprise* was the first vehicle built at the Palmdale plant in Southern California and was delivered to NASA in 1979. It was initially named the Constitution, but after a nationwide renaming program, undoubtedly due to the popular TV program *Star Trek*, the shuttle was renamed. It was never intended to go to space but instead was used for testing and evaluation. Two years later, the *Columbia* was launched.

The *Challenger*, the third built, flew in 1985 and was used more than any of the other operational orbiters. Of the 135 launches of the five, only two were destroyed. There were two tragic platform crashes that resulted in major loss of life. The other four shuttles are now on display. One resides in the Air and Space Museum, and the other three are in museums in various states.

# CHAPTER 2

The Disaster

Crossing the bridge that faithful day was stop and go, as usual. The thousands of drivers crossing the bridge, all going various places, were oblivious to my need to get to work on time. Trying to forget the traffic, I thought about the evolution of the US space program. After Robert Goddard, the father of rockets, developed space boosters and made it possible to reach space, missiles and rockets became the norm.

Starting in 1962 when the Russian Sputnik was launched, American astronaut John Glenn launched into orbit. Following the Redstone rocket, Thor missiles were implemented, and then our international continental ballistic missiles were brought on board by the US Air Force. In the late fifties, NASA became the preeminent space organization.

The Mercury and Gemini programs eventually led to the Apollo flights and the moon landings. Ironically as a basic trainee in 1963 going through boot camp in the Air Force, I volunteered for a special assignment. Little did I know they were doing human studies on space flights in preparation for the Mercury program.

Years later, while still in the air force, I was an air crew member on the C-141 aircraft. We transported the crew of the space shuttle *Columbia*, who took part in a space flight in a specially built module back to the United States. The airtight module housed the three astronauts after they splashed down in the Pacific Ocean, near Guam.

The crew had to stay in the module all the way to the Houston Space Flight Center in Texas.

The three astronauts were in quarantine after landing back on Earth after three space walks in orbit. During the flight home, we could see and talk to them through the portholes in the modules but couldn't have physical contact. The quarantine worked. Years later, while at Lockheed, my training team and I were singled out by STS Mission 28. We were acknowledged for our support in a classified Department of Defense launch.

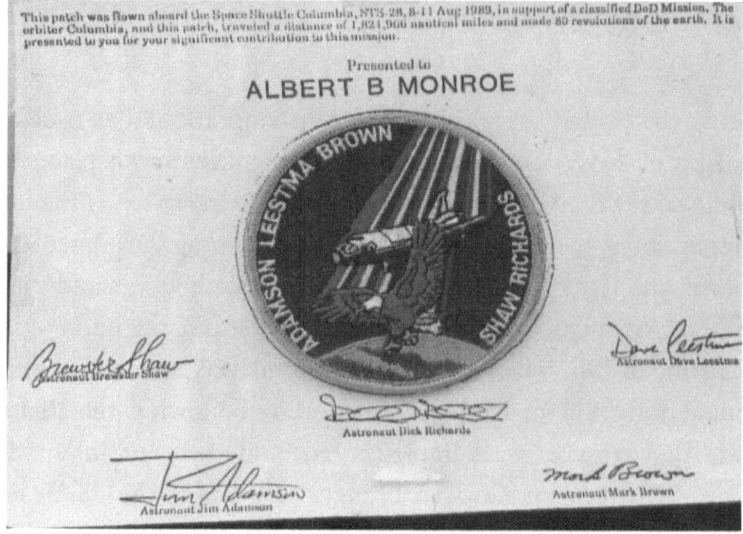

Awarded for my Participation in this mission

By the time I reached work that day, the parking lot was almost full. I was both excited and nervous but never envisioned the scenario about to unfold. One of my coworkers on the same training team pulled in next to me, and he too had a cautious look on his face, which made me worry. We chatted for five minutes before heading in to work

We could sense the atmosphere as soon as we walked through the doors, which made me feel strange. Looking at my staff member's faces, I could tell no one slept the night before, and like me, they were anxious to get to work. The normal greetings and office pleasantries weren't there. I could feel the tension and uneasiness all around me. The whole training team gathered in the classroom to watch the launch.

We were still over an hour away from the countdown at Kennedy Space Center in Cape Canaveral. It was odd to see weather this bad in Florida. Since we had such a beautiful day in Sunnyvale, we never gave any thought to the weather down at the Cape. A cold wave had gone through Florida, dropping the temperature to a no-fly zone. Ice had formed on the skin of the orbiter. At that point, the ground crew should have notified Mission Control, but whatever reason nothing was done. The launch should have been delayed or scrubbed. The ground crew should have insisted, which would have caused a twenty-four-hour delay. Nobody wanted this. More delays. Human pressure overruled safety that day. A giant mistake that led to loss of life.

Of all things, freezing temperatures in Florida were the culprit that terrible day. You would think that heavy rains, a hurricane, or lightning would've stopped the countdown. The flight director in charge gives the go or no-go decision. That person's decisions are final, and all the responsibility of the launch falls on his or her shoulders. The entire world was watching. After months of delays, the pressure to launch became another factor for the tragic mishap. Time passes so slowly when you're waiting for something to happen. Feeling somewhat hungry, I searched out a glazed chocolate donut and coffee to fill the void after being too anxious to eat prior to leaving for the office.

Looking at everyone in the room, I could see my whole training staff were anxious and nervous. Down at the launch site, the seven astronauts were finally strapped into their seats in the tilted-inclined position for takeoff. Sitting in that inclined position for hours puts a huge strain on the human body. Finally, the word was given. After a recheck with all the system directors, which included electrical, hydraulics, and so many others, the thumbs-up was eventually given for the countdown to begin. Now it was only minutes away, before our space shuttle's twenty-sixth launch into space with the first civilian on board.

Large crowds attended each launch, and this was no different. For this significant launch, though, about three hundred guests were present, along with thousands of spectators who had parked their cars around the roads and highways to view the event. Thousands more contractors in most of the fifty states were watching intensely as well. With the increased pressure from Washington, several dignitaries, an astronaut's family members, support staff at the Cape, and congressional folks from DC took the 'elevator up to the mission control center roof for a clear view of the launch.

Mission Control Center

On board for this mission was a special astronaut: a civilian School Teacher, Ms. Christa McAuliffe. Her function was mission specialist. Our country envisioned ordinary folks of all occupations could participate and work in space. The government decided to assess this theory, and after a big campaign, Christa was selected. NASA was hesitant but gave in. She trained for over a year, just like any other astronaut going into space. Anticipation was on everyone's face. Surely every school in the nation was curious and anxiously watching as this event had been well advertised in advance. Our government was counting on the potential impact this mission could have on the number of kids who wanted to become engineers and scientists, especially girls. Prior to Christa, the only American woman in space was Sally Ride. NASA now has women as mission specialists, mission commanders, and copilots.

Lieutenant Colonel Ellison Onizuka, the first Asian American astronaut, was also on board as a mission specialist. He carried a soccer ball with him to orbit, signed by all the members of his daughter's soccer team. Ironically, the ball was eventually returned to the high school thirty years later. In 1994, Sunnyvale Air Force Station, which housed the Blue Cube was renamed to Onizuka Air Force Base in his honor.

Everyone is ready. With a go, a thumbs up from all the systems, the flight Director started the count-down. Everyone's anticipation is now at their highest level. At T-minus 10, the mechanical arm atop of the shuttle moved away. At T-minus six, one of three main engines came alive. At T-minus five, the second of the three engines ignited, and at T-minus three the third engine fired. Now, at T-minus two, the huge solid rocket boosters (SRB) ignited. Now with all three liquid oxygen engines burning and the two SRBs firing, the *Challenger* lifted off pad 39B, on its way to orbit.

Not there for the launch, but after experiencing other launches, I could feel what many were going through watching the launch. The noise and vibration of liftoff is horrendous. The whole area shook from the pressure created by the five thrusts of the engines. The mission control building, only two miles away from the pad, also can feel the launch despite its distance. Typically, the orbiter is thrust upward and ascends into space ninety miles down range and at an altitude of eighty miles up. Usually, the two solid rocket boosters burn approximately 212 seconds, run out of fuel, and are jettisoned away. Attached to parachutes, both tanks gently float back to Earth and land in the Atlantic Ocean. A large salvage company retrieves them and tows them back to the Kennedy Space Center, where they are refurbished and scheduled to be used again. Another one of NASA's secrets—the SRBs were initially built to reduce costs and were supposed to be reused three or four times, but just three have been reused out of nearly three hundred launches.

As the vehicle rise off the pad, and clears the tower, command and control are turned over to the Houston flight director and his staff for the remainder of the mission. The Houston team of controllers monitor every phase of the launch. The status and data of the launch of each system is relayed to the Houston flight director, and they use this input to make live vehicle decisions. Computers on the vehicle are programmed to operate each phase of the flight all the way to orbit, and both the onboard crew and ground control can change the program and command the spacecraft. In the first couple of launches, the Houston team was reluctant to use the computers for command and control, but it worked flawlessly. It became so common that everything was computerized, and most of the time every position monitoring the launch just watches their screen for any anomaly.

Back at the training center in Sunnyvale, everyone was jubilant, as the launch went off smoothly. With the three orbiter engines at full blast and the two SRBs creating a whopping 6.5 million pounds of thrust, it powered the *Challenger* into space. We feel relieved, but that changed immediately. What happened next changed the course of our space program. Joy changed to sorrow in seconds. In a matter of seconds, the whole STS program came crashing down, not only affecting us, in Sunnyvale, but in all fifty states.

Back in Houston, the flight director gave the signal. Only seventy-three seconds into the flight, the ground screens went blank. There was no data coming in. Each screen showed red, indicating a malfunction. It caught everyone by surprise. Nothing like this had ever happened before. Large telescopes closely monitor every launch, and the word came back that there had been a large explosion. Everyone was dumfounded. The flight director asked for data, but there were none. Word from the telescope team indicated there were four or five streams of debris floating back down to earth. It was now certain that it was a major disaster. The shuttle had blown up, and there were no survivors. Just three minutes, and it's all over. The flight director immediately closed the command center, not allowing any press to enter.

It looked like a Fourth of July fireworks display. White plumes of smoke were streaming back to Earth. Debris was spread out for hundreds of miles. The large crowds was in disbelief. The dignitaries were confused. I could just imagine how the million of contractors felt watching all over the country. Many knew immediately felt their jobs were in jeopardy. At the Cape, the salvage ship that was normally used to retrieve the SRBs after they were jettisoned was now told to look for survivors. There were none. A large-scale rescue effort began to collect any debris. They even brought in the navy's deep-sea submersible rescue vehicle to search at deeper depths. The word went out to the public for any boat, ship, or vessel who found any debris to please turn it in. It took days to recover the seven souls on board and weeks and months to gather anything to be able to make an accurate picture of what had just happened.

Back in Sunnyvale, we couldn't believe our eyes. We were all in shock and total disbelief. All the preparation and training for the *Challenger* and crew were gone in minutes. I imagined all the schoolteachers and school kids watching in every state who must be devastated. Mary, one of my coworkers on the training team, was crying furiously and had to be consoled. Two months earlier when Mary had done a specialized training class at Cape Canaveral, Christa McAuliffe was one of her students. Mary had been impressed by Christa's knowledge, and the two became friends. When Mary saw the explosion, she knew the fate of Christa and the other six on board. The ground crew at the Kennedy Space Center hoped there were survivors, but at that altitude and speed, no one could survive such an explosion.

In Houston, everyone was scrambling to find out *what happened.* Analyzing the data, when the signal to power up was given, it showed one of the SRBs O-rings had failed, allowing the liquid oxygen fuel to seep into the SRB, which had caused the enormous explosion. No one on board had a chance at all. The O-ring is a rubberized seal, and due to the wintry weather and the long wait, it had shrunk slightly, losing its tight seal, which caused the two fuels to mix. After months of inspections Morton Thiokol, the SRB manufacturer, was eventually blamed for the failure. But the wintry weather was also a culprit. The Kennedy Space Center ground crew speculated that at warmer temperatures; the seal may not have shrunk to the point of failure. Not only were seven astronauts lost and the *Challenger* destroyed, but there were several medical and biological experiments on board.

The loss was devastating to NASA. They'd had so much riding on this launch. The nation went into mourning. Ronald Reagan, then the president of the US, went on TV and tried to console the nation. Many of our allies were not as enthusiastic as they previously had been to do space studies. The sixty-four-thousand-dollar question was: *What's next?*

Congresspeople, other than those from Florida, talked about canceling the program. Week after week, new theories arose as to just what had happened. Every nook and cranny was inspected. Finally, it was decided the O-rings and the cold temperatures were the cause. The flight director also took some of the blame, although all vehicle indications had shown normal readings. His judgment was questioned. But overall NASA's space shuttle program survived, and within a year, NASA was back in business.

It took close to four months of searching at sea to retrieve *Challenger* parts from the ocean. The shuttle was painstakingly pieced back together and inspected thoroughly and repeatedly, but nothing was found—just the O-rings. Morton Thiokol redesigned their parts for NASA. In Houston, they upgraded their computers to better detect malfunctions and give early warnings. At the Cape, they redesigned the crew compartment with safety measures.

NASA was determined not to let something like this ever happen again. But anything manufactured is vulnerable to breakdowns, and all one can hope for is that it won't break down on your watch. All the updates and improvements worked. Going to space is a dangerous business, and there is no guarantee. Loss of lives and vehicles is inevitable. Any potential astronaut entering training is aware of the consequences. Once the launches were resumed over the next twenty years there were no incidents, until *Columbia*. All the systems worked flawlessly putting satellites, including the Hubble Space Telescope, TDRS, and other significant payloads, in orbit. But no one expected another major disaster.

# CHAPTER 3

The Aftermath

It's hard to bury seven astronauts. With almost fifty states involved with the shuttle, everyone was affected. NASA did not want this accident to ever repeat itself, and with all four space centers—Marshall, Goddard, Houston, and the Kennedy Space Center—along with the major contractors, they worked furiously to get the program going again. Problems in the Middle East were on the rise, which caused a shortfall in the budget. NASA could not get any more money from Congress, and it had to make do. Morton Thiokol in Utah, the SRB builder, redesigned and strengthened their tanks to better standards. After months of testing, it was determined they were space worthy again.

In Sunnyvale, the loss hit us hard. We knew there would be six to eight months of inspections and delays, just like before. That terrible day, no one was able to do work, and our supervisor Joe reluctantly let us leave at lunch time. Weeks and months went by fast, and eventually the shuttle seminar was postponed until the launches resumed. Even with the delay, all those satellites had to be launched. With the shuttles on hold, our only recourse was to use unmanned vehicles.

South of the Bay Area, near Bakersfield, Vandenberg Air Force Base had two launch pads that were mainly used to launch DOD payloads aboard the smaller rockets, like the Atlas and Delta. The Vandenberg site was smaller than the Kennedy Space Center. During the 1980s, Vandenberg AFB was thought of as another NASA shuttle launch site, but funding was never approved. Today, Vandenberg still is used for the less powerful rockets, like the Delta rockets.

With the space shuttle program on hold, we knew the launch site at Vandenberg Air Force Base would soon get busy. It was used to launch satellites from unmanned missiles like the Delta, Atlas, and Titan rockets. Our space shuttle course did not touch on rockets and missiles, so we had to do research on these types. My boss, Mr. Redd, myself, and another team member traveled to Vandenberg to get the lay of the land. We inspected the two pads, did as much research as we could, and picked up the specs on the Atlas and Titan vehicles. Soon after that visit, my team was tasked with developing a new course on unmanned launch vehicles.

My boss and the rest of the training team were determined to beat the deadline we were given. We worked furiously on this project and working long days and on weekends was common. Mr. Redd had worked for NASA in the early days and knew the shuttles in and out, but he didn't know much about rockets and missiles. Fortunately, with the Vandenberg documentation our training package slowly took shape, but we needed a rocket expert.

Reading specifications, looking at schematics, and analyzing blueprints are great, but we were still struggling. We put out a request for a rocket expert. Several months went by, and in walked a breath of fresh air. Bob was the man—the expert we were seeking. He had the expertise to decipher all the documentation we were struggling to comprehend. Within weeks the course was back on pace. We were happy, and the workload seemed to get easier. But nothing lasts forever. Bob was from Florida, and he had never been west of Houston. It was his first time in California.

On October 17, 1989, a 7.2 earthquake hit the Bay Area. Hundreds of people were killed or injured. So many were displaced. Bob had never experienced anything like this, and it scared him out of his wits. While having dinner that evening, in his underwear, the shaking started and grew stronger. He ran outside, forgetting his door key. He had to call the building manager to get back in. It shook him up to the point he put a transfer back to Houston and was gone within two months.

Thankfully, we had gained a lot of experience in his brief time with us and could complete the course. Mr. Redd, saw the course coming together and relaxed somewhat, and eventually calmed down. He was a stickler on time. If it took overtime to finish it, you'd better do it. My other team members gave him a nickname: the old country boy. He was a tall older gentleman, very elegant. He wore thousand-dollar, custom-made suits every day and drove the biggest Cadillac I had ever seen. He had everything monogramed with his initials. At times it was challenging to collaborate with him, but we worked it out. With the course finally completed, Mr. Redd assumed his other passion. Every weekend, and even on weekdays, he would drive to San Francisco and put on flower arrangement shows. He was good at it and was sought out by celebrities, fashion shows, and other events. He told me he made more money doing that than at his Lockheed position.

My Shuttle Course Trainers

Back at the Blue Cube, mission controllers and their staffs were scrambling to keep the low-flying satellites in orbit. With the shuttle program on hold and the unmanned rockets not being utilized to their capacities, a backlog developed. Something had to be done soon. All satellites will eventually slow down. They decay over time in orbit and must be replaced. The ground controllers try to stay ahead by keeping their batteries fully charged, but over time the solar arrays on board just won't keep them functioning properly. Gradually, the low-flying satellites will slowly die out and crash back to earth. When that happens, there's nothing anybody can do but just watch the spectacular reentry and fireball. The highflying satellites have a much longer lifespan and are not exposed to the same gravity pull as the low flyers.

In the Palmdale plant in Southern California where the shuttles are manufactured, all the recent changes since the *Challenger* crash were incorporated in the latest shuttle being built. The engineers were trying to install duplicate and even triplicate redundant systems in case a failsafe system was needed. One major upgrade was that all the electrical wiring was double wrapped to prevent heat and moisture buildup.

A major issue NASA needed to resolve were the shuttle's heat shields, sometimes referred to as tiles. During any shuttle reentry back into the earth's atmosphere, all vehicles heat up tremendously to over fifteen thousand degrees, which is a critical phase of the flight. The tiles are attached to the bottom surfaces of the shuttle to protect the vehicle during reentry. The heat shields are basically glued on. Several types of glue were developed to adhere the shield to the vehicle. For a while they were having problems keeping the tiles attached. Rain could sleep in and cause a breakdown of the glue. However, the tiles on the *Columbia* vehicle were already operational and deemed to be safe. The new glue replaced some of the tiles, but not all of them—something that came back to haunt NASA years later.

At Cape Canaveral, safety was paramount. A large crew escape system was developed in case the astronauts had to rapidly exit the vehicle's crew compartment before a launch. It looked like a shark tank cage that was attached to cables where the crew jumped into the cage and slid down using cables all the way to the ground. Even the orbiter crew compartment was modified with all doors opening outward to expedite exits. NASA had learned a hard lesson a few years earlier when two astronauts were killed in a fire when the crew doors opened inward and could not open in time.

On a visit to the Cape, my boss and I went down to view a launch. Up at the top where the astronauts enter the shuttle crew compartment, we were offered to evaluate the escape system. You could see the whole coast of Florida that high up on the launch pad. My boss declined, stating it would interfere with our tight schedule. I was glad he refused because it did look very scary being that high up.

Over two hundred major modifications were made to improve the shuttles and the program. Another significant modification included changing the flight suit the astronauts wore during the launch. Prior to the accident, the crew just wore flight jump suits, which offered little protection other than being fire resistant. After the crash, a whole new suit was developed to give them protection in an oxygen-rich cabin environment. A pressure suit was initiated and worn at launch and reentry, which gave them more protection.

NASA eventually resumed their astronaut training program, actively seeking women and minorities to be mission specialists, pilots, and copilots. Even new mission control staff personnel were hired. Back in the eighties, NASA had only recruited military personnel into their program, but this was a new era. Even our allies were not excluded. Citizens from Israel, Japan, Great Britain, Germany, and Canada applied. Scientists, engineers, and mathematicians were recruited from the major schools around the country.

All NASA facilities and their other smaller departments in the forty-eight states were tasked with upgrading. Essential contractors, like us in Sunnyvale, and others were asked to do the same. NASA wanted a total makeover of everything, thinking it would be the answer, and it was for another twenty years. But nothing is failsafe. In the aftermath of the *Challenger* accident, NASA learned many lessons and made thousands of changes to improve their program. It was a total effort and eventually the shuttle program resumed, putting many significant satellites into space. Everything seemed to be working well until another tragic day almost twenty years later.

After the crash in the late eighties, business was booming. The economy had really picked up. A lot of new technological breakthroughs were implemented. Millions and billions of dollars were made worldwide, and much of that money was being hidden in overseas accounts. Dot. com companies came on the scene and generated a *lot* of money. Around the world, companies and whole countries got rich practically overnight. Big oil was the cause. Massive oil field reserves were found in Saudi Arabia, Kuwait, Iran, and others, which generated a tremendous amount of wealth. At that time, we were allies with Iran, Turkey, Saudi Arabia, and a few other countries. It appears Saudi Arabia had the biggest oil reserves, and Iran had the power in the region due to their military might and close ties to us. But the Middle East started to unravel. In the late eighties, many Middle Eastern countries were modernizing and updating their infrastructure. But you can't change old customs and ideologies overnight. Heads of states tried but failed since half their populations were older and didn't want to change mindsets. Money helps, but changing ideology takes years—if possible, at all.

The United States was the leader of the free world and head of the United Nations. We were looked at as protectors. To be protectors we had to stay abreast of what was going on in the region. Iraq, Yemen, and others had oil, but not as much as Saudi Arabia. If you don't have enough, how do you get more? One way is to take it, which Iraq tried by invading Kuwait. The United States provided Kuwait assistance and quelled the invasion. The Gulf War began. Communication and surveillance were critical. Using old technology like landline telephones, telegram, and teletype will only get you so far. The answer for this new kind of war was in satellites, which could monitor, listen, and do surveillance.

CHAPTER 4

Shuttle Capacity

Let me tell you another NASA secret. When we look at a shuttle, it looks like an ordinary airplane and does fly like one in Earth's atmosphere. It was made to be reusable and land like a regular airliner. It has distinctive features like an airplane when it's in Earth's atmosphere, but in space, it doesn't operate like a plane at all. At launch the three main engines in a vertical position fire first; then the solid rocket boosters produce thrust, which gives the spacecraft its main boost, lifting the craft off the pad into orbit.

The two boosters and liquid fuel eventually run out. In fact, the boosters are depleted in about two minutes. The large bright-orange tank in the middle holds the liquid hydrogen and liquid oxygen fuel, which power the three main engines. After about eight minutes, the tank runs out of fuel. This renders the three engines useless, and the huge tank is jettisoned away. But, unlike the big orange tank, the SRBs fall back into the Atlantic Ocean, are retrieved, and can be reused. The last push to get into orbit is made by the two smaller orbital engines. The craft enters orbit approximately at 28,000 kilometers an hour. Circling the earth takes ninety minutes. Now in orbit, cooling the spacecraft is critical. An hour after launch, the indoor side of the cargo bay doors must be opened to cool the spacecraft and allow the computers and other systems to operate properly.

Unlike a typical airplane, the spacecraft flies backward and upside down while in orbit to protect the front end of the vehicle. There is a lot of debris—or space junk, as NASA calls it—flying around that could collide with and damage the spacecraft. It could be a tool or piece of equipment lost on another mission, like a screwdriver, pair of pliers, or pieces of satellites that did not fully burn up. At twenty-eight thousand miles an hour, these objects are moving fast and can pierce the shuttle. The craft is now in a low-Earth orbit, anywhere from 150 to 500 kilometers high. You would think our engineers would have figured out a way to discard the three main engines along with the middle tanks to reduce the overall weight, which would allow more cargo to be carried.

The shuttles were constructed to be super gliders once in Earth's atmosphere. Upon entry at approximately two hundred thousand feet up, the pilot must maneuver to glide the shuttle to its landing. The onboard computers must be accurate to land back at Kennedy Space Center (KSC). The craft cannot glide forever and must land without power, so the pilots practice pitching, yawing, and rolling the spacecraft to increase or decrease speeds. The craft is exceptionally light, as it's mainly empty, having dropped most of its weight. Upon entry, depending on its speed, it can't go far. There is no fuel to power the three main engines, so it cannot maneuver around thunderstorms, lightning, high winds, or severe weather.

The orbiters were built in sections and put together like no other aircraft. The main section—the part that resembles an airplane—was manufactured to be reusable multiple times with excellent maintenance, and other parts, like the SRBs, could also be used again. The middle oxygen tank is never reused after a flight. Over a thirty-year career, and with 135 space launches, the *Discovery*, *Explorer*, and *Atlantis* launched well over thirty times each. The *Challenger* and *Columbia* were destroyed in flight but had several flights each.

The craft, about the size of a Boeing 737, is sixty-seven meters long and sixteen meters across. The fuselage is comprised of the front, the middle, and the aft. The front, or forward, houses the nose section, a reaction control system, small thrusters, and the crew compartment. There is an upper crew compartment, called the flight deck, where the pilot and copilot sit, which contains three other seats, and the lower section, where the food galley, toilet, and sleep quarters are and where the mission specialists sit. In the lower section, computer racks and other critical systems are found. There's no gravity in space, so members of the crew just float from the upper to lower areas. Crew members do not require oxygen in the crew compartment but transferring to the cargo bay through the oxygen chamber requires a pressurized space suit.

The mid-fuselage section holds the wings and the cargo, or payload, bay. The wings, just like those on an airplane, has ailerons for maneuverability during landing. The cargo bay is the meat of the shuttle. About the size of a school bus, it has many features, and can accommodate several experimental modules while in space. Once on orbit the cargo bay doors are opened, rendering the cargo bay unpressurized. Crew members working in the cargo bay must wear protective suits. Satellites are launched out of the cargo bay. It also houses other equipment, modules, space experiments, or parts like those taken up to construct the International Space Station (ISS), which is still flying today. A sixty-foot Canadian robotic mechanical arm is attached on the right side of the fuselage to aid in grabbing and putting payloads into orbit. On the bottom of the fuselage heat shields or tiles are attached to protect the space craft on deorbit. Over twenty-seven thousand tiles are attached to the orbiter to take the massive heat buildup and protect the vehicle during reentry.

In the aft part of the vehicle (lower back) is where the vertical stabilizer, flaps, and rudder are located, which are used mainly for landing. The back of the craft also contains the orbiter maneuver system (OMS) which is used to power the vehicle while in orbit. These smaller engines can be gimbled, which means they can be moved up, down, right, or left, which can change the orientation of the spacecraft. They have their own internal fuel cylinders that house the fuel. Eight smaller thrusters located near the flaps are used for maneuverability. The OMS engines are small and shoot out short burst's seconds at a time—not only to change the orientation but moving the craft to a higher orbit— and are used in docking with the International Space Station.

While in orbit, the mission's work commences. Satellites that need to be launched are prepared and checked out. If docking with the ISS is required, the onboard computers take control, placing the orbiter in proximity near the shuttle to mate up. One of the pilots usually takes over and assures the two crafts are locked together. Maintaining pressurization is critical. Without it, a major disaster could occur. Once docked, now both ISS and shuttle doors can be opened to allow access from one craft to the other. If satellites are to be placed in orbit, the mission specialists don their pressure suits and transfer from the crew compartment into the cargo bay.

Depending on the payload's size, weight is not a factor because there is no gravity in space, so everything weighs the same. Bulk or a bigger satellite may hamper a crew member's ability to maneuver the payload around. Some of the payloads are flimsy, so careful handling is required. If anything happens, the Canadian robotic arm is used to grab the payload and place it properly. The arm can assist the astronauts when they're doing a spacewalk or working outside of the station. A fix platform can be attached to the arm, allowing the space walkers to stand or work freely. On a particular mission a payload already in orbit needed to be retrieved. The arm grabbed the payload and brought it into the cargo bay for an equipment change. After the work was completed, the satellite was placed back into orbit using the arm. The Hubble telescope was repaired this way. The arm brought the telescope close to the shuttle, and new lenses were exchanged, keeping the Hubble functioning for years to come.

In the 1990s, NASA envisioned using the orbiters to go back to the moon and even beyond, but those ideas were short lived—probably due to cost and NASA deciding to change their approach. Congress ordered a more practical use. The unmanned rockets were used to launch payloads into orbit. Also, two commercial companies came on board and were being used to launch payloads into space. By then the Russians had their own space lab, MIR, already in orbit but near its end. The US also launched some of its payloads atop the Russian rockets out of Ukraine.

Our major allies, along with other nations, wanted to build a research vehicle where astronauts from any country could work in space. Funds were allocated and the International Space Station came into being. The shuttle got a new life. With so many countries footing the bill, funds were no issue. This was a major undertaking. Never had something this massive ever been tried. Everything had to fit exactly, and there could be no mistakes.

The major parts were launched in sections and mated to each other. Then the smaller pieces were brought up and secured in place. Due to the station's larger sections, the shuttle was the only way to transport them into orbit. Along with our shuttles, commercial launches, and the Russian heavy unmanned rockets, the ISS parts arrived and were then assembled in orbit—a great engineering feat! Well over fifty combined launches placed two hundred sections in space, and it slowly grew, piece by piece; it was like a giant Lego set being constructed in space. Though it took years to build, it is fully operational today. The ISS flies over your town every so often. Currently onboard are four American astronauts, a Russian cosmonaut, and mission specialists from Israel and Japan. More astronauts will be switched out in 2022.

CHAPTER 5

Cradle to Grave

NASA space shuttles are unique, but nothing lasts forever. The program lasted thirty years, and if it weren't for the huge cost, overruns, delays, and budget cutbacks, I feel the shuttle transportation system (STS) would still be in existence. There were six shuttles built, all of them in the Rockwell Palmdale plant in Southern California. It was a manufacturing marvel with so many contractors participating. The original concept was a space vehicle that could do anything in space, and at one time thought of as a way for man to return to the moon. Some even envisioned going to other planets and into the far reaches of the Solar System. Our government was feuding with the Soviet Union for space supremacy, and this led to many mistakes.

After President Kennedy's speech about landing on the moon the race to space escalated, but to achieve those goals, a crewed vehicle was thought of as the answer. Using unmanned and crewed rockets to put payloads into space have limited capabilities even though today's launches are safer than those years ago. But a shuttle controlled by a crew can do much more.

The path to orbit took so many steps along the way. After the craft is built at the Palmdale plant, the next big step is to ensure the spacecraft is airworthy. The shuttles, too big for wind-tunnel tests, go through a series of fly-bys. NASA checks the orbital dynamics, making sure the craft can fly and land once it enters Earth's atmosphere.

Using a modified Boeing 747, the shuttle is placed atop and taken up to approximately sixty thousand feet and is eventually detached. The shuttle, once released, becomes essentially a large glider like what they experience in space. Then the orbiters go through a series of tests, making sure all airplane systems, like the flaps, rudders, ailerons, and other systems, are functioning properly. Once all the airworthiness tests are passed, the shuttles are certified and rolled out for public display before their initial trips down to the Kennedy Space Center.

Getting it to KSC is no easy task. The spacecraft is about the size of a Boeing 737 airplane. It can't be trucked, shipped, or flown to Florida. Remember: on Earth, without jet engines, the shuttles cannot fly like a traditional airplane. So, there's only one way.

NASA flies it to Cape Canaveral on top of a modified Boeing 747, called the *shuttle carrier aircraft* (SCA), while others refer to it as the *Mother Ship*. The shuttles, being empty and without any fuel sources, like the solid rocket booster (SRB) and liquid oxygen tank, is placed on top of the SCA with special bolt-down attachments. For safety reasons, the big Boeing makes several stops along the way for rest and refueling. It is a slow process, and the trip takes approximately three days.

Shuttle Carrier Aircraft

This is the start of the space vehicle's incredible journey, one that can last well over thirty years. Once it arrives in Cape Canaveral, Florida, the shuttle goes through hundreds of steps before it becomes operational. Landing at the Cape, the Mother Ship with its payload atop taxis over to the orbital processing facility (OPF). This is a large maintenance facility, a part of the shuttle complex. This large facility, sometimes called a hangar, can accommodate two shuttles simultaneously.

At the OPF, all the orbiter's maintenance is done. Inside the OPF a special rig was built called the *Mate-De-Mate Device*, used to lift and lower the space vehicles—essentially, it's a large crane. The giant crane is attached to the orbiter, where it is lifted from the big Boeing. With enough space, the crane slowly lowers the shuttle to the ground. Now the real work begins.

For the next few months, the shuttle goes through extensive checks in which every system is put through series of tests and then certified again. Multiple tests with numerous systems operating all together are thoroughly analyzed. This initial step in the OPF is critical. With redundancy for every system, it is a grueling, thorough inspection. Breakdowns in orbit can jeopardize the safety of the crew and the vehicle. In space, small breakdowns can easily lead to larger problems and possible catastrophes.

The next several steps are just as critical. After months of testing, inspections, and evaluations and finally being certified, the orbiter is moved again. The vehicle is then taken over to the largest building at the KSC, the vehicle assembly building (VAB) for mating together the SRBs and the big orange liquid-oxygen tank. The VAB structure is huge and stands about thirty-eight stories high. It's hard to describe its massive size without seeing it. Touring it for the first time I thought, *What a marvel of engineering.*

Another oddity: NASA maintenance ground personnel say the VAB has its own weather system near the ceiling. The VAB also has the four largest doors of any building built in the world. Before it became the VAB, it was constructed for the Apollo spacecraft and the big Saturn five rockets.

Vertical Assembly Building

In 1966 it was designated the vehicle assembly building once the shuttle came online. It was the only facility that is used to mate the sections that make up the shuttle. Some payloads, which must remain in a vertical orientation, and even classified and DOD payloads, can be inserted into the orbiter's cargo bay, while in the VAB, away from prying eyes. The VAB has also been used as backdrops in seven blockbuster Hollywood movies, including *Moonraker*, *Apollo 13*, and five others.

First, the bottom sections of the two SRBs are brought in—one for each side. Next, the second SRB section is stacked on the first, and then the third piece is added. With fuel already inside each SRB, this is a delicate procedure. Any spark and an engineer might be in heaven earlier than expected. All the sections are connected and mated together. An extensive inspection checks all the lines, valves, and attachment points and is now ready to mate with the huge orange liquid-oxygen and hydrogen tank.

Now the liquid fuel tank is ready to be mated. The empty tank is driven in to the VAB on a huge platform in a horizontal position. The large bridge cranes align themselves over the tank, and the cables are attached lifting the orange tank to the vertical position. Moving ever so slowly, the crane repositions the tank between the two stacked SRBs. Still attached to the crane, the SRB and tank are attached to each other. All the fuel lines are connected, and once again a thorough inspection takes place. Once the ground personnel are satisfied, they certify their work. Now all that's left is the shuttle itself.

Depending on the launch schedule, this whole mating process can take weeks, or even months. Approximately two weeks before a scheduled launch, the shuttle is towed over to the VAB, where the bridge cranes lift the orbiter to vertical and positions it up to the mated SRB and oxygen tank. It is slowly maneuvered into position then mated to both propulsion sources. All attachment lines and fuel lines are carefully connected. You know what happens next. Extensive tests, inspections, and certifications ensure the spacecraft is fully operational and ready to fly.

All that is left is to move the assembled shuttle out of the VAB to the launch pad. Another critical step in its journey. How is this done? With another engineering marvel. While all this work is ongoing at the Cape, preparing the vehicle for space flight, astronauts three thousand miles across country at the Houston Space Center are going through extensive training themselves. We'll get back to the Cape soon.

CHAPTER 6

Astronauts Training

Every shuttle mission is different. All five orbiters go through the same exact procedures to get to space, but out of the 135 launches, only a few had the same objectives. The astronauts, on each mission, depending on their function and position, are fully trained. The pilots ensure the shuttle makes it to orbit and maneuvers the space vehicle for reentry and landing. Five mission specialists make up the rest of the crew personnel on each mission. They are usually the scientists, chemists, medical folks, and the engineers. Performing all these duties in space is no easy task and is the culmination of weeks, months, and even years of training in their areas of specialty. In the eighties and nineties, astronaut training was different than it is now.

In the early years, the International Space Station (ISS) did not exist, so only after the space transportation system (STS) program ended in 2011 did the training program change. After the original astronauts were selected, like the Navy's Alan Shepperd and Neil Armstrong and five others prior to the STS program, did NASA open its astronaut's training program. Up until the late eighties, the STS program only selected military pilots who had a certain amount of flight experience and had to meet other stringent requirements. Once shuttle launches became routine and various experiments could be studied in the weightless environment of space did NASA opened its training program to civilians.

Many applied, but only a select few were admitted. Initially, a bachelor's degree was required, but those with expertise and PhD's in their specialties were even considered. Other requirements, like character and being a team member, were sought after. Physical requirements were paramount. Past ailments or histories of certain issues like heart attacks, strokes, and other medical concerns were carefully judged to make it to the next stage of the selection process. After US Senator John Glenn, one of our original astronauts back in the sixties in the Mercury and early Apollo days, went up, in his mid-seventies, anyone could apply. NASA's qualifying restrictions were less stringent. Once NASA envisioned using the shuttle to go to the moon and beyond, ordinary folks were allowed to apply.

Selected candidates, after passing all the stringent physical restraints, go through an initial two-year training program. The trainees all go through the same type of training at the same time. Set up as a seven-man team, they go through a grueling and rigorous ordeal. Training is not just eight hours a day; one's whole life is consumed with daily training, meetings, workshops, and studying at night, which leaves little time for family.

Another dirty secret that NASA often neglects to mention is that there is a high divorce rate among the STS program crew. With five major NASA centers spread out across the country, there are always required training sessions at these various locations. NASA also doesn't tell its recruits that once training is completed and they become fully certified shuttle astronauts, there is no guarantee they will ever go to space.

Waiting to go to space is frustrating. After years of training, it may be many more years before being assigned to a mission. Astronauts are essentially government employees, and it's not a profession that pays six figures. In fact, new trainees are paid half as much. There is notoriety and prestige in being an astronaut, who keeps some employed, but many get disillusioned and fed up with all this training and leave NASA for the more lucrative private contractors looking for this expertise. Once the STS program was cancelled in 2011, NASA cut its workforce almost in half, including nearly one hundred astronauts.

So just what type of training is required? Entering the initial two-year program, all sorts of orbital dynamics are taught. Working and living in a weightless environment is required, so much of the training involves weightlessness. But how do you achieve this while fighting gravity here on Earth? NASA built a large swimming pool called the neutral buoyancy lab. In their space suits, astronauts are submerged twenty feet down in the tank. Then they can experience zero gravity while underwater. With safety scuba divers assisting, the training is usually in two- or three-hour sessions. This is one of the most critical training sessions and requires a great deal of stamina.

Learning to operate the sixty-foot remote mechanical Canadian robotic arm is required for the mission specialist. One of their duties may be launching satellites out of the shuttle's cargo bay. NASA had a mock-up in which this training takes place. Without gravity, weight is not a factor, but the size or bulk of the satellites may be awkward to maneuver, requiring special handling. The arm is so sensitive that just a slight touch on the control button may be too much. It was like a kid's video game, but with expensive consequences if a payload is damaged.

Launching a satellite using the Remote Canadian Robotic Arm

Payload in Orbit

NASA recruited their pilots from the military. All of them came in with numerous flight hours in various aircraft. NASA decided all crew members be pilots for safety reasons and required their mission specialists to be certified pilots as well. The trainees are trained in the Air Force T-38, the same jets used today for the air force pilot training course. The NASA pilots had to go through glider training to make sure they were efficient in landing the spacecraft once back in the Earth's atmosphere.

One of the modifications NASA made to improve safety was to develop the emergency egress system (ESS). It is now a critical task that's required for all astronauts. After the fire that killed three of our astronauts and following the *Challenger* disaster, this emergency system was created. Consisting of four-gauge baskets, the system allows one to slide down a long cable from the upper entry point to the ground before takeoff and outside the space vehicle. It is like a zip line. These baskets hold two or three crew members at a time. On a tour of the pad, my boss and I declined an invitation to ride down. It was just too scary.

Working on Shuttle engines

Human endurance training was also required. It was envisioned that astronauts could spend weeks, months and even years in the weightlessness of space. What effects does it have on the body? NASA wanted to establish some baseline requirements just to qualify to be a crew member. Centrifuge training evaluated the endurance capabilities of the human body. In the early days, before the STS came online, the Mercury and Apollo programs used the centrifuge to run tests. Ironically, early on in my military career I volunteered for a training session but did not know what was involved. I never went back.

Once the civilian mission specialists completed their two-year program, they could only simulate their training on the ground and needed to be in orbit. Long-term stays in space can be harmful. If we were ever able to establish a permanent base on the moon, we would have to know the effects on the body first. Even those astronauts returning from six months on the ISS have suffered effects from their long endeavor.

Learning to walk in space was a critical task. Many mission specialists volunteered, but only a select few were chosen for this training. The ability to fully use the capability of the STS program and what humans can do in space has increased tremendously. Satellites already in orbit that are damaged or need replacement parts can now be fixed in space.

If the payload won't come to you, you go to it. Space walks are required to go to the payload and do the necessary repairs. In the early days astronauts had a tough time doing a spacewalk. Sweating was common, which fogged up their headgear. Holding on to tools like pliers and screwdrivers was difficult due to the size of the space suit gloves. Various pieces of equipment have been lost on space walks and are still flying around. Maintaining the temperature and oxygen flow in the suit was initially difficult, but years of practice, new upgrades, and modifications to the suit finally made it possible for repairs to be more routine than before. Tethered walks ensured the astronauts would not be unattached at any point, but this proved cumbersome, and jet packs on the back of the suit allowed for maneuvering around without the tether line.

New Suit for EVA's

In the nineties and aughts, NASA changed its training program again. This program still exists today, even after the STS program was shut down. The training program is used for the crew member on the crewed rockets. Now the program is divided into five areas. The first one is the neutral buoyancy lab for learning to live and work in space. This is the underwater training that teaches future astronauts how to maneuver in space.

The second phase is becoming proficient on the Canadian robotic arm. The third phase is learning everything about the International Space Station (ISS), still flying today. The ISS is the only spacecraft in orbit, where astronauts from any country can do their weightless studies. The fourth phase is critical learning to speak Russian. US satellites and crews are launched atop Russian rockets. Russian equipment and systems are in Russian. The language is required to communicate with the ground. NASA puts all their recruits through a Russian language course that they must pass.

As I mentioned before, the fifth phase of training is pilot training. Other types of training include the multiple axis training (MAT) and environment staffed unit (EMU)—basically the space suit. The MAT is like a seat that swivels and swings in different directions. This puts enormous strains on the body, which affects blood flow. Astronauts are launched in a small capsule and can lose their orientation, and the crew members can experience rolls, spins, and wobbling affects. EMU training involves all aspects of the flight suit. Today's suits are sophisticated, with various attachments available depending on the mission.

After the STS program was cancelled, Astronauts were launched aboard the various rockets, Not knowing where they might land, various scenarios were considered. Other types of training—like water and land survival—are required. Landing in the ocean in a small capsule on a windy day may lead to parachute entanglement, sinking, or the chutes being pulled for miles across the ocean, so water survival is essential. Believe me, it is not as simple as it looks. The parachutes used to slow the descent and land in the water are extremely heavy. If the chute happens to land on top of the capsule, it may be able to float for a while, but it will eventually sink, and all insides could drown. It is critical to escape the modules as soon as landing completes. I was lucky enough to go through this water survival course in my military days.

The Russians use a different approach. They land their capsules on land, mainly in the Siberian Desert. There are benefits and consequences of doing both. Landing in the ocean may lead to drowning, whereas landing on land may lead to sudden impacts if the parachutes become detached. Landing in extremely inclement weather in the winter months is another detriment. Temperature below freezing may cause equipment to freeze. Escape hatches may also freeze, which creates another hazard. Being on land with nothing but snow around may lead to hypothermia. You may think these scenarios would never happen, but nothing is guaranteed to work properly. Every astronaut must be initiative-taking rather than reactive when it may be too late.

Computer training is another sore subject. Every crewmember must also be computer proficient, along with knowing their own specific duties or tasks, With back-ups to the back-up, there are many redundant systems crew members are required to know. As new computer programs come online, there is so much classroom computer training. Most of this type of training is boring, but it is essential for safety purposes. It cannot be guaranteed that every system will operate properly, so everyone must train for contingencies.

Now, with most of the basic training complete, the astronauts are ready to go to space. Constantly training on their specific tasks, they anxiously await their assignments. Going to space is the culmination of all their efforts, and that's when they *truly* become an astronaut.

With the crews trained at the Houston site, the shuttles are almost ready. When we left chapter 5 to see how the crews are trained, the orbiter was still in the vertical assembly building three thousand miles away at the Kennedy Space Center. Let's return to KSC and see how the vehicle makes it to the launch pad.

CHAPTER 7

Moving Shuttle to Pad

When we left chapter 5 the shuttle, completely assembled in the VAB, was ready to be moved to one of the pads—either 39A or 39B—for launch. Both pads are identical and mirror each other. Prior to the program shutdown, 39B was more active. Today 39A is used more than the other for the big Saturn rockets.

Just how do we move the shuttle? NASA uses a massive crawler transporter to move the vehicle to the pads. The shuttles, with SRBs and oxygen tank attached, along with the mobile launch platform (MLP), is set on top of the crawler. At eighteen million pounds, it is no easy task moving that much weight. Built in the sixties, it was another marvelous engineering feat.

A team of designers and engineers from the ALCO company, a locomotive company, with a low bid of approximately fourteen million dollars, won the contract to build the vehicle. ALCO's team developed the massive crawler. It was originally built to move unmanned rockets, like the massive Saturn booster that took us to the moon. After the shuttle program came online, and during the thirty years of the STS program, the crawler has transported every orbiter out to the launch pad. The crawler transporters have been around for over five decades and are still going strong. NASA expects to continue to use it, no matter what vehicles require transport.

Shuttle on Pad

The mobile transporter is massive. About the size of a major league baseball infield diamond, it is 139 feet long, 114 feet wide, and weighs approximately six million pounds. It is powered by six large, 16-cylinder diesel engines, with two huge generators on each end. These diesel engines, overhauled and with new parts, have kept the transporter functioning at optimal performance for over fifty years—another NASA gem!

Consisting of eight roller tracks, like those found on Army tanks, each track has fifty-seven shoes or 456 shoes total. NASA claims each shoe is one ton. Two were built early in 1965. After major testing, inspections, and certifications, both crawlers became operational in December of 1966. And in August of 1967, the crawler moved the Apollo Saturn Five Rocket to pad 39B for the first time.

To keep the crawler running, there are a host of system and subsystems that power the transporter. It can run both on AC and DC power. It has both hydraulic and pneumatic power. This hydraulic system has an equalization jacking and leveling system, called the JEL, that keeps the jacks and platform level and in sink. Because the pads are built on a sloping upgrade, the transporter use its screw-jacks to maintain this level orientation as it moves up the five-degree incline to the pad. Massive belts are used to move the screw jacks. To ensure the belts run smoothly, they are automatically lubricated and monitored by the onboard computers. The minimum height is twenty feet and can go as high as twenty-six feet. With all this power there's a reason they call it a crawler. It moves at a whopping speed of one mile per hour loaded and two miles an hour unloaded. The trip out to the pad is at a snail's pace. The average time to travel the five miles is approximately eight to ten hours.

At eighteen million pounds, moving this much weight is a monumental task. There are few roads that can manage that big of a load. NASA had to develop and construct a road strong enough to accommodate this amount of weight. But like most things, NASA got it done. The five-mile roadway the crawler moves over to the pads was specially built. Traveling over a normal road, the weight would cause cracks and sinking. Special layers of limestone, gravel, crushed stone, sand, and dirt are compacted, down nine feet to accommodate such a heavy load, and the road is constantly resurfaced.

Using diesel fuel, it takes five thousand gallons for a fill up. Consumption is terrible, as one gallon is used for every thirty-two feet— approximately 165 feet per gallon. A little bit worse than a Lamborghini. Just like the ongoing training at Houston Space Flight Center, the crawler training is extensive. Crawler driver training is intense, taking twelve to sixteen months to be certified.

There are two crew cabs, one on each end, for steering. The crawler never turns around. Once is gets to its location; the driver goes to the cab on the other end. Both cabs are extremely insignificant compared to the size of the transporter. The steering wheel is only sixteen inches in diameter, but when turned it moves large gears and uses a computer laser-guided detection system to steer the vehicle. The system is so accurate it can take the load out to the pad and set it down within a half inch of its goal.

To monitor the transporter systems while in route, a whole building facility was built. All functions were computerized, so alerts, bells, and whistles are used if any errors or malfunctions are detected on the crawler end route out to the pad. While in route, the crawler may be automatically stopped if any computer, either the one on the transporter or in the building complex, senses an out-of-range alert level.

Now with the STS program shut down, the transporter is used to move the large Saturn rockets to launch pad 39A. The other pad, 39B has been disassembled, leaving 39A the only launch site at the Kennedy Space Center. NASA decided only one pad was needed for future launches, which reduced operational costs. While not a secret, NASA launched both historical missions—the Apollo moon rocket and its last shuttle launch in 2011 at pad 39A.

In chapter 5, I mentioned the steps the space vehicles go through on its journey. Still in the VAB, the huge overhead crane cables are attached to the mobile launch platform (MLP), and both structures are lifted enough to allow the crawler to be positioned beneath and in place. The cranes gently lower the whole assembly on top of the transporter. Secured by bolts, the complete structure weighs close to eighteen million pounds and is now ready for its trip.

Once again everything is inspected and certified. The next big step is the trip to the pad. With a *go* signal from the manager, the complete transporter assembly slowly leaves the VAB and heads out to the pad. A team of engineers in the building complex and another twenty or so ground maintenance personnel accompany the crawler in its journey. With everyone on headset communicating with each other, the trip can be immediately stopped if anything out of the normal is detected. As mentioned earlier, if a red alert is issued, the crawler is automatically halted until the problem is fixed.

Almost at the pad, the crawler moves up the five-degree incline next to the fixed service structure (FSS) and place the MLP on supports directly over the exhaust flame trenches. The empty crawler makes the trip back at two miles an hour, taking half as much time as the initial trip. The MLP has cutouts for the SRB exhaust on both sides and a larger cutout for the three main engines. At liftoff, six million pounds of thrust is vented down through the MLP cut outs and out through the flame trenches. With the space vehicle on the pad, another structure called the rotating service structure (RSS) can now rotate around the shuttle and close it off from both the elements and for protection.

The FSS has tubing and cabling for water lines and fuel loading and electrical cables for communication and general maintenance. An elevator is used to take the crew up to the top for crew entry. A beanie-type cap is moved over the top of the shuttle to vent the escaping oxygen for safety. At the top of the FSS, right across the platform, there is a small room called the *white room,* where the astronauts go through their final preparations. Originally built for the Apollo program, all the structures were painted white. At the top of the FSS near the crew entry hatch there is a porcelain toilet. Tour guides tell their audience that it's the last toilet on Earth.

With everything in place, the RSS can rotate and maneuver around to protect the space vehicle while it waits for its launch, which may be up to three weeks later. Three reasons were given as to why the RSS was built. In the Apollo days, there was no rotating service structure, just the fixed service structure. One reason is that payloads can now be inserted in the cargo bay at this point. Protection from the Florida elements like rain, hurricanes, and lightning can do damage if severe enough. Too much rain may cause electrical problems. And the third reason is for maintenance and crew access.

Takeoff may take weeks after the craft is on the pad, so the RSS, rotated in place, keeps the shuttle safe. Every system is checked and double-checked. All the ancillary systems, like the liquid oxygen and nitrogen fuel, are stored in two tanks near the launch site. The sound suppression system and the MLP locking bolt are inspected and certified. With all systems online, all that's left is for a mission go.

The sound suppression system is critical during initial liftoff. NASA used water to suppress the sound and keep the noise and vibration to minimum levels. The water is stored in a large tower tank near the pad. Water floods the MLP to reduce and deflect sound waves from damaging the shuttle and payloads. Two seconds before the SRBs start, the water is pumped under pressure smothering the MLP and flame trenches. And at six million pounds of thrust structures could melt if not cooled.

Another ancillary system is made up of the explosive bolts that hold the shuttle upright while it is on the pad. High winds may topple the shuttle if it's not secured. To do this, four explosive bolts attach the SRBs to the MLP on each side. The locking bolts hold the space vehicle in a vertical position and prevent it from tilting, which can be a problem at liftoff. Once both SRBs and shuttle engines are firing, the bolts are explosively unlocked, allowing the vehicle to ascend. It is essential they unlock simultaneously to keep aligned vertically. It is not known what would happen if one or two bolts remained locked. I suspect at that amount of thrust being exhausted it would just disintegrate the bolt.

Five hours prior to launch, the liquid oxygen and liquid nitrogen are loaded in the big orange tank that fuels the three main engines at takeoff. The tanks are kept very cold and are carefully monitored. The liquid oxygen is loaded first. Then the nitrogen tank that sits atop the oxygen tank is filled. This is a very delicate process, and safety is paramount. Any lightning strikes, sparks, or any types of fire are not allowed. No maintenance type of service is allowed while fueling is taking place. A dropped piece of equipment may cause a spark that could ignite the fuel, and you just bought the farm. The fuel is kept in cylinder tanks, and oxygen and nitrogen are on either side near the launch site and channeled through pipes through the FSS to the space vehicle. Any lengthy delays and the fuel is drained back into their storage tanks for temperature control.

Florida is the lightning capital of the world. Our pads are located near Cocoa Beach on the east coast, close to the Atlantic Ocean. It rains almost daily in that region of the country, and lightning strikes are prevalent. To protect the shuttle while it's on the pad, three large towers were erected at each pad. The three towers that surround the pad have a cable attached between them to absorb or deflect any strikes away from the shuttle to keep it safe. Every shuttle launch occurred in mild weather, but if there are delays and inclement weather—like a thunderstorm with lightning—and fuel is still in the orange tank, this could be a major problem.

With the shuttle ready to go, all that's left is the crew. With a *go* signal, the crew is transported to the pad approximately two hours prior to launch. The crew take the elevator up to the top level of the FSS and into the white room for final preparation. After final checks on each crew member, the seven astronauts take the few steps over to the vehicle, enter the shuttle hatch, and take their seats. The seats are in the inclined position, so the ground crew assists each crew member with a safety harness, helmet, microphone, and any other need. With the crew ready, all systems are powered up, and communication is established between the flight director in the control center and the crew on board.

At this point, the beanie cap is moved back. All the communication, fuel, and cables are unattached and moved away. The RSS takes twenty-eight minutes to rotate out of the way, exposing the shuttle. With all support equipment stored, the ground crew secures the shuttle hatch, takes the elevator down to the ground, boards the bus, and are driven away from the pad. Waiting for the countdown is intense. Over at the control center, the flight director gets a thumbs-up signal from each system. With all systems a go and no red alerts, the countdown begins. If there's a problem before the SRB ignites, the countdown is halted, and the crew can exit the shuttle by deploying the EEES, where the six baskets carry the crew away from the pad and down to the ground.

The countdown to launch is the last step of the shuttle's journey before orbit. At this point everyone—the crew, the mission flight director, each system director, both the maintenance and ground personnel, and trainers like myself—are nervous and holding their breath. I still remember that fateful day the *Challenger* disaster occurred after just approximately two minutes of flight. My team and I at the Sunnyvale training center three thousand miles away in California were extremely nervous and worried that day before launch. After hearing about the freezing weather at the launch site, we were even more concerned. Years of preparation and training were gone in two minutes. It was a devastating loss.

For a successful launch, a sequence of events must occur. Any deviation before launch could result in a disaster. The timeline is essential. The computers at the control center and those computers onboard the shuttle take over the launch. With a go signal from all systems, the control center director begins the countdown right at twenty seconds. At eight seconds, the oxygen and nitrogen deflectors to reduce explosions are ignited, and the three main engines are ignited at six seconds. At two seconds, both SRBs ignite, and the sound suppression system floods the launch pad with copious amounts of water. Simultaneously with the SRB ignition, the explosive bolts unlock, and the space vehicle gently lifts off the pad. Clearing the FSS, command of the shuttle is now turned over to the Houston Space center for the remainder of the flight. The shuttle begins a slow roll, down range. This reduces stress on the vehicle. After two minutes both SRBs, now out of fuel, are jettisoned away and land intact in the ocean so they can be used again.

Almost in orbit, the oxygen/nitrogen fuel is depleted in approximately eight minutes. The tank is jettisoned, and with the shuttle up so high the tanks burn up on reentry. The three main engines are useless throughout the flight after they are out of fuel. To reach final orbit, the two orbital engines are used to place the vehicle in orbit. The shuttle maneuvers to an upside-down position with the rear of the shuttle flying backward. After thirty minutes in orbit the shuttle bay cargo doors must be opened to cool down all systems.

Now in orbit, the crew can say they have reached true astronaut status. The ground flight director now goes to the mission schedule. What the mission calls for is the next step in the journey. If a satellite is to be launched, the mission specialists get ready to do their thing.

CHAPTER 8

Working on Orbit— Launching Payloads

The next three steps after the shuttle reaches orbit are critical. The steps are working in orbit, deorbit and reentry, and landing. It may not sound important, but so many things can go wrong during these three phases. And we have already seen what could go wrong. The space shuttle *Columbia* went through this during reentry, when it was destroyed with the tragic loss of the spacecraft and crew in 2011.

With the space vehicle traveling at twenty-one thousand miles an hour and circling the globe every ninety minutes, all systems must be running to keep the mission on schedule. Oxygen, communication, and computer systems are essential. If problems arise in those areas, the mission will be scrubbed, and the vehicle will deorbit and land back on Earth. Various systems, like the water purification system, must work and are critical to the continuation of the flight.

Most flights have several objectives. One is doing weightless studies; another may be to launch satellites or Department of Defense (DOD) payloads. Docking and supporting the International Space Station (ISS) may be another scenario. A space walk, called an EVA, might be scheduled or repairs to satellites already in orbit. Being in low orbit, approximately 150 to 200 miles up, most work is done in this range. On occasion, the mission may require a higher orbit. If a higher orbit is needed, the two orbital engines, along with the smaller jets are used to push the vehicle up. With 135 launches, most of the flights did not encounter any problems, accomplishing their full mission schedules. But there were a few in which the mission was canceled, resulting in an immediate return to Earth.

The schedule dictates what is to happen next. The crew is anxious to get started, but depending on the mission, the crew might have to wait for certain timelines—like launching a satellite at a certain time. Satellites are sensitive and placing them in orbit is critical to their lives. The space vehicle transitions from darkness where it may be zero degrees or below to being in the sunlight where temperatures may be in the hundreds of degrees every forty-five minutes. A part of the astronaut training includes using these timelines. This part of the training occurred at the Sunnyvale training center, where my team and I were instrumental in providing this essential work.

The payloads are mainly developed by commercial companies who are anxious to get their satellites in orbit after years of development and testing. Most of them have a team of developers, engineers, and monitors to take over control once the payload leaves the shuttle's cargo bay. Some may require a higher orbit and will use jets to boost them to a different orbit.

Let's look at how satellites are launched. It sounds simple but is no easy task. As mentioned, launching satellites are critical, and a checklist is used to ensure every step is in sequence. Most satellites are super expensive, and if one is damaged someone may lose his or her job. Some are as small as a basketball, while others are massive, like the Hubble Space Telescope. The sixty-foot Canadian robotic remote arm, attached to the shuttle cargo bay, is used to assist grabbing, and placing payloads out of the cargo bay into orbit. There is no gravity, so all things weigh the same, but overseeing and maneuvering large payloads may be cumbersome, especially with the space gloves astronaut's wear.

Before anything can happen outside the orbiter, the mission specialists that are responsible for launching payloads must be prepared. With the cargo bay doors open for cooling, outside in the cargo bay the mission specialist endures the environment of space. The crew must be protected, so donning the space suit is required. While inside the pressurized crew area, they suit up, checking all the suit systems, like oxygen, water cooling, and communication. They exit the crew compartment through the hatch into the chamber, transferring them to a weightless environment; then they exit the chamber into the cargo bay, where they can begin their duties. Once again, timelines come into play. If in darkness, they may have to wait awhile for sunlight to warm the payload. If the payload is too big for the mission specialist to manage, the robotic arm may be deployed. A crewmember inside the crew compartment controls the arm while looking out the window. With everyone in communication, this is normally a straightforward process.

With the shuttle in orbit, the cargo bay doors open, and the mission specialist is doing his or her job—working and launching satellites. Let's examine some of the other important duties, like an extra vehicular activity (EVA), the name for a space walk away from the shuttle and docking with the ESS or the now-extinct Russian MIRS space lab and our space lab that no longer exists.

CHAPTER 9

Working in Space—EVA

Since our first American astronaut, Alan Shepherd, went into space for the first time in 1961 in a fifteen-minute sub orbit, we have been protecting our crews from the harshness of space with a specially designed space suit. Staying in the pressurized, oxygenated capsule or space vehicle where it is safe, an astronaut can work in regular clothes. Step outside without a suit and you would die in seconds. In the early Apollo days of space travel, moving around in a space suit was a cumbersome task. NASA knew they needed a suit to protect the astronauts when they landed and walked on the Moon.

ISS023E021088

Working in Space

Years later when the STS program came online, launching satellites became the norm. Crews in orbit realized launching anything mechanically out of the cargo bay was risky, so staffed crews were needed for this operation—but they had to be protected. Once a crew member leaves the pressurized comfort of the crew cabin, he or she is exposed to the environments and dangers of space. NASA named it an extra-vehicular activity, or EVA. We just call it a spacewalk. Another NASA secret was the reluctance of mission specialists to be away from the shuttle in the openness of space. But it soon became evident that the astronauts sometimes needed to be outside the cargo bay to successfully launch payloads.

Before the shuttle came online, unmanned rockets mechanically launched their payloads, often with issues, and several payloads were lost. NASA thought that launching smaller satellites was often more problematic than launching bigger ones. The astronauts were tethered to the cargo bay to prevent flying off into space and being lost. Being tethered got in the way and hindered launching the larger-sized payloads out of the cargo bay.

NASA had to invent a better system. A group of engineers worked on this problem, and several iterations of the suit were produced. The first spacewalk took place in 1966 in preparation for men to walk on the moon. Everything had to work successfully. Like parachutes, it had to work the first time. You just can't bring it back and get another one. The improvements continued for years. New packages were developed, and the suit went through major changes, like the detachable jet packs, on-suit oxygen systems, communication packages, and anti-fogging now allowed unlimited capabilities. The improvements and updates now reduce NASA fears about space walks.

These specialized packages allowed the mission specialists to be unattached from the spacecraft and maneuver around. Initially, helmet fog-up and suit cooling, along with a host of trivial things prevented the suit from working properly. The overall bulk of the suit made it extremely tiring trying to work in the weightless environment of space. Many of the space walks, scheduled for approximately four hours, were cut short due to astronauts experiencing sweating and nausea. Without using the jet packs while away from the orbiter, one solution was using the Canadian robotic arm, which aided the astronauts in their walks away from the orbiter. The arm is equipped with a foot-attachment plate the crew can stand on to keep them upright and stable.

So much effort went into the suit training and learning how to maneuver in space. The large swimming tank, used for buoyancy training, tried to prepare the crews at home on Earth. Simulation training is great and helps to prepare astronauts but being in the actual weightless environment cannot be simulated back at home. Certain problems like fogging up and using the jet packs could not be duplicated while in the tank.

With new versions of the suit, training became paramount. Every time modifications or enhancements were developed, the crews had to learn how to use and operate those apparatus. Suit training became constant. As the suits become lighter and more adaptable for humans, the training became much easier. The EVA training team is such a critical part of the astronaut two-year program. Being protected from the harsh environment of space is essential and using new suit improvements keeps the crewmembers safe and allows them to perform tasks more quickly, efficiently, and easily than ever before. At home, you can simulate training in tanks, but working in space is on-the-job training.

EVAs took a major turn in the early nineties, when it was determined astronauts on space walks were a critical skill needed and required to do the maintenance on the space station, which was carried to space in pieces. The Canadian arm, with numerous EVAs, aided in assembling these large sections. Once the larger segments was assembled, the mission specialists must attach the smaller sections together for alignment and strength. EVAs got so important that NASA established an EVA office where all aspects of the suit design, working in orbit, and suit repairs came under their umbrella. A task force of engineers was established to work on issues, like suit design, using various materials, steel over aluminum, and others. For dexterity, special rubberized gloves replaced the prior ones. The gloves were even heated, which gave mobility in the hands and fingers.

In the early 2000s, before program cancellation, NASA envisioned using the shuttle to go to the moon and beyond. This could take years, so learning how to make repairs in orbit was essential and the only recourse. Anything manufactured doesn't last forever, which includes the ISS. It needs periodic repairs. Ordinary tools may not work in space, so special tools had to be created and carried to orbit to manage any issues. With a lot of debris flying around in space, junk may hit and damage the ISS and could punch holes in the support structures. Sharp edges could rip the astronauts' gloves while working. Large tears or rips to the gloves could be devastating, oxygen could be leaked from the suit, and cause loss of life. You just can't simply return to Earth, so all repairs must be made in orbit.

With the shuttle transportation system program canceled, space walks are still required to repair the space station and make updates to extend the lifespan of those payloads already in orbit. The lifespan of the station is limited, and in a few years, without modifications, upgrades, and replacement parts, the International Space Station will reach the end of its life. There has been talk of overhauling the ISS to extend its life to 2030. Now with private companies with their deep pockets—like SpaceX—and more involvement from countries worldwide, there may now be enough much-needed funding. Countries like the UK, Japan, Canada, and Germany can help continue ISS operations past 2030, or we may use another type of space module for exploration and living and working in space. We'll see!

Using the Jet Pack from propuslsion

Working in Space -Launching a Payload

CHAPTER 10

Leaving Space

With all tasks completed, it is now time to come home. Astronauts say that being in space is going to the final frontier. But like Dorothy said in *The Wizard of Oz*, there's no place like home. Dorothy clicked her heels three times and was home in seconds. To leave orbit, there are three critical phases, which take approximately two hours to complete. Chapter 9 talked about working in space, but in chapter 10 I cover how astronauts get back home. It is not as easy as you think. Leaving space is critical and everything must be exact, or the space vehicle may miss its intended reentry target and run into a host of issues, all leading up to failure. You just can't point the nose down and descend back to Earth. At launch the orbiter ascends straight up. It doesn't work like that coming down.

With everything secured, the cargo bay doors closed and locked, all equipment tied down, and the crew back in the suits, they wait for the signal to start the deorbit process. The seven crew members are back in their original seats, with the two pilots up front, three mission specialists sitting behind them, and the other two below in the lower crew compartment. Without the fuel, the tanks, and payloads at launch, the orbiter weighs close to five hundred thousand pounds compared to the six million pounds it started with at launch.

With a go signal from the Houston flight controller to deorbit, the shuttle is halfway around the world, traveling approximately 28,000 miles per hour, The craft must slow down for deorbit. Remember, for the entire flight the shuttle has been flying upside down and tail first for safety reasons. The space vehicle uses the Earth's rotation to get into position to land at the Kennedy Space Center. Energy is the key element in landing. Using it to your advantage by bleeding off or increasing speed is the goal. This is accomplished by increasing or decreasing the angle of attack, or *bank angle* as it's often called. So many technical terms are used, but for our purposes I will give you the laymen's terms. Prior to entering the Earth's atmosphere, the orbiter is a space vehicle, then after Earth reentry it flies like a regular airplane.

When everything is ready, the onboard computers maneuver the shuttle in proper alignment to begin a series of deorbit burns. The computer fires the OMS engine approximately three minutes, which slows the vehicle down. The burn must be exact. The vehicle enters a fixed corridor. Inside this corridor, being too high or too low may be devastating, and it may be over before you know it. It takes about twenty-five minutes for the shuttle to descend and reach the Earth's upper atmosphere. Then the RCS system, those small jet engines in both the forward and aft sections, fire to position the nose of the craft upward to 40 percent. This is a crucial maneuver. You want to expose the bottom tiles under the wings and fuselage to absorb the massive heat. Looking at pictures of reentry, you see a reddish-orange glow around the shuttle. This is from the extensive heat build-up. The temperature rises inside the orbiter, but the suits with their internal cooling help protect the crew.

Now the shuttle is falling rapidly with the nose up. The remainder of the RCS fuel is burned off for safety before entering the extreme heat during deorbit. Slowed down to 17,000 miles per hour, the orbiter enters its most critical stage. Falling rapidly, extensive heat is built up. This heat and now friction raise the molecules surrounding the vehicle to rise to 3,000 degrees or more. No craft could stand that much heat and would start to melt and be destroyed. The heat shield protects the forward area, underside, and other areas from damage. The tiles must work as advertised and protect the shuttle from the massive heat building up as it descends.

Preparation for Leaving Orbit

NASA had key issues with the tile in the past. The heat shields kept falling off. The glue did not adhere to the skin of the vehicle when exposed to water and caused the heat shields to become unattached, exposing the metal skin to this massive heat during reentry. This is devastating to any mission. And as we saw with the shuttle *Columbia* in 2011 on de-orbit and reentry, the missing shields destroyed the *Columbia* with seven crew members perishing. NASA immediately solved the problem, and the 28,000 shields have done the job, without any further issues.

Just what material make up the heat shields? For the underside and wings, they used reinforced carbon. The upper fuselage uses a high-temperature protective covering. Nomex blankets cover the cargo bay doors and many other areas. The rest of the orbiter is covered with a low-temperature insulation.

The shuttle starts to maneuver into position to go through reentry and line up for landing. With the nose still up 40 degrees, she continues to descend rapidly. At this point, another critical part of the vehicle's mission is blackout. During the extensive heat build-up and ionization, no communication can get through, so all communication with the ground is lost. This lasts for about twelve minutes. This blackout, as it is called, keeps the Houston ground crew in suspense, not knowing the status of the vehicle as it rapidly descends. Hopefully, the high-powered telescopes can pick up the shuttle and notify the ground crew that there are no issues.

After deorbiting, next is reentry into Earth's atmosphere, where communication can be reestablished, and the shuttle can now fly like a normal airplane. Unlike a traditional aircraft, the shuttle has no fuel to power the three engines and is now essentially a large glider. But this large craft cannot go extremely far. The onboard computer takes over the controls and flies the shuttle. The computer turns the shuttle into a series of S-turn banking maneuvers, which slows its descent and decreases speed. At this point the computers aligns the shuttle for its approach back to KSC. A radio beacon aids the landing and alerts the crew they are on course.

Approximately twenty-five miles out and at eight thousand miles high, the shuttle commander takes over for landing. The commander flies the shuttle through a series of "S" these S turns, turning left to decrease or increase speed to reduce the velocity. Then at 1,700 feet she enters terminal energy alignment management, called TEAM. These are basically circular patterns that help line up the runway and manage speed. To decrease speed even further, the nose is set at minus 20 degrees, where the nose is slightly pointing down, for the approach, which is a lot lower than commercial airliners.

With the shuttle in position and on course, she is ready for touchdown. At 2,200 feet above ground, the commander pulls the nose up to slow the rate of descent, and the landing gear is deployed. The speed brake on the tail may be activated to help decrease the speed. At 225 miles per hour, she touches down. To aid slowing the vehicle, a large parachute in the tail section is deployed, along with the speed break, also a part of the tail section, which may be partially or fully activated, depending on speed, to increase drag, slowing the vehicle substantially. Eventually she slows and comes to a stop. Mission over—well almost!

Back at home the crew is anxious to deplane and walk on Earth. But there is still a short wait. The shuttle onboard computer programs must be shut down, so the crew deactivates most switches, and during these twenty or so minutes, the orbiter has cooled enough, and any obnoxious gases have dissipated. Now the crew can deplane. The ground crew is there to assist if needed. Being in the weightless environment of space for days or weeks leads to loss of strength in the torso and legs. Some crew members have needed support to walk immediately upon return. I heard some crew members were so happy to be back at home they knelt and kissed the ground.

Mission over? Not yet. The crew is taken back where they are debriefed to make sure any issues or problems that occurred are immediately looked at and dealt with. At this point, the mission is over. The crews are given plenty of time off to relax and regroup, be with their families, and allow their bodies to be re-exposed to gravity on Earth. Depending on the schedule, another mission is set for liftoff with a new crew and new agenda in a few weeks.

The orbiter is towed back to the orbital processing facility, where it is extensively inspected. Latest updates or packages may be included at this point. The tiles are checked and may need replacements. In a few months, with everything certified, she is ready to fly again. It could be months or even a year or more before the shuttle goes up again. Most of our operational shuttles have been reused many times over—some up to twenty times. But the program was halted prematurely due to costs, overruns, funding, and cutbacks.

I talked about the cradle to grave concept, and this is the cycle the shuttle goes through. In the next few chapters, I'll cover the *Columbia* disaster, one of two orbiters destroyed in NASA's thirty-year history. Next, we'll look at the Hubble Space Telescope, still in orbit, and some of the significant satellites that have been launched.

CHAPTER 11

The Columbia Disaster

In the last few chapters, we found out how the space vehicles are developed, prepared, launched, and operated in space. We looked at how we deorbit from space and land back at home. I took you through the life cycle of the STS space shuttle program and its multi-purpose uses, from putting payloads in orbit and doing scientific research to supporting the International Space Station. With 135 launches, all five vehicles performed brilliantly, with only two losses. The two losses: (1) the *Challenger* explosion, which occurred on January 28. 1986. Seven crew members were lost when it exploded after only two minutes of flight. And (2) the *Columbia* disaster on February 1, 2003, in which another seven astronauts perished during reentry, when the shuttle broke apart and crashed. Some experts say the *Columbia* crew may have been saved with a space walk over to a rescue vehicle. But this did not happen and resulted in the loss of the crew and the *Columbia* orbiter as well.

Americans are the most ingenious people in the world. We have developed and built amazing structures, crafts, and machines. But anything manufactured is not 100 percent indestructible and can be destroyed with the right circumstances or events. Humans cannot make anything failsafe. We can come close, but when dealing with Mother Nature, she wins every time. We keep pressing our luck to no avail. The German airship the *Hindenburg* was thought to be indestructible, but on its approach for landing, wet lanyards used to secure the blimp were struck by lightning, causing a massive explosion, and downing the craft. And many of us saw the movie *Titanic*. Once again on a craft thought to be indestructible, an iceberg ripped a big hole in the side of the ship, and within two hours she sunk, causing massive loss of life. These are only two, but you can probably think of a few more.

Going into space is dangerous, and the crews do not worry about what could happen. Being trained and prepared when their number is called is their main objective. Between the United States and Russia there have been several crashes, disasters, and loss of life. This is a volunteer force, and there are so many inherent risks that they just try to be prepared and hope everything works as advertised. All systems are redundant and even have back-ups to the back-ups, but things can and do happen, as we have seen. You mourn for the loss but keep going forward. Even after both the *Challenger* and the *Columbia* disasters, none of the astronauts in training quit NASA, and everyone stayed together. When disasters do happen, we examine why, try to piece together the scenario to determine the cause, make the adjustments, and move on.

Finally, it was here: launch day. After three years of preparation and training, twelve different delays, STS-107 was ready to go. All members of the seven-person crew were ready to become official astronauts. It was January 16, 2003. I had left Lockheed Missile and Space Company after ten years and moved to Las Vegas, Nevada. Since I'd served twenty-two years in the military and thirteen years in the tech industry, I thought I was ready to retire. I thought that moving to Vegas would give me that opportunity to retire early and travel, but I was wrong.

I got bored early on, so I decided to pursue another one of my passions in life: teaching. I had a degree in education and taught others while in the military and at Lockheed, so training and teaching were already in my blood. I started out as a part-time teacher, teaching at both elementary and college levels. Then, I worked as a full-time teacher and became a professor at a private college for the next thirteen years before finally retiring for good.

The space transportation system (STS) program and keeping up with the various launches was also in my blood, so after leaving Lockheed I stayed abreast of NASA and the shuttle operations. Several times on vacations and trips to Florida, I would coordinate my trip with an ex-coworker to meet and see another shuttle launch. Seeing a launch is something special that everyone should experience—you'll never forget it.

Back in Florida, STS-107 was scheduled as a routine mission—nothing NASA would say was special or out of the ordinary. No space walks or satellite launches were on the agenda, just a mission to do routine scientific experiments in orbit. The time schedule in space seemed excessive. Sixteen days in space was longer than most flights; staying an average of seven or eight days was more typical, but this extra time was needed for the experiments to thrive and plants to germinate. Also, with specially built scientific labs in the cargo bay, the mission specialists could work without the need of space suits.

Everything was a go. I wasn't there for that launch, but I was told the crew was anxious to get going after all those delays. The crew was well seasoned with thirty-six months of training under their belts. All of them were in their forties, and if this mission was scrubbed, some may never get another chance to go to space. Rick Husband, an air force colonel, was the commander. He had previous shuttle experience, having been the pilot on STS-96, an ISS support mission. The pilot was navy commander William McCool. Ironically, William's parents live in Las Vegas, less than two miles from me.

Col. William McCool

Five mission specialists were a part of the crew, including Llano Ramon, an Israeli air force colonel and pilot. He was the first Israeli person to go to space. Mission specialist Kalpana Chawla, an aerospace engineer, was from India and had prior experience on a previous flight. The third mission specialist was navy captain David Brown, a doctor who was trained as a flight surgeon. It was his first flight. Laurel Blair Salton Clark, also a navy captain, trained as a flight surgeon as well, who worked on scientific experiments. The remaining crew member was Lieutenant Colonel Michael P. Anderson, the payload commander, who was trained as a physicist. With so much expertise on board, NASA had a lot of confidence that once the shuttle was in orbit things would go as planned.

Columbia STS-107 went off flawlessly with thunderous applause from the crowd who had shown up to watch the launch. Once the shuttle cleared the tower, control turned over to the Houston flight control center. On course, all systems were in the green. After two minutes, the two solid rocket boosters were jettisoned and parachuted down, landing in the Atlantic Ocean. Now, the *Columbia's* three main engines were powered by the huge oxygen and nitrogen orange tank. It would be another eight minutes before that large tank exhausted all its fuel.

With almost a minute left, a large piece of foam insulation on the oxygen tank came off and struck the *Columbia's* leading edge of the left wing. Nothing unusual. On most flights, tiles and pieces of insulation are ripped off. Both leading wings are covered with the high-resistance heat shields where protection is most needed during reentry. On the ground, the high-powered cameras picked up a flash high on the oxygen tank that shot downward and collided with the left wing. Out of fuel, the oxygen tank was jettisoned away and eventually burned up in the atmosphere. There was no indication of system malfunction, and everything was still in the green. Once they reached orbit, the crew prepared the craft for its upcoming schedule.

Back at Houston, the initial pictures from the camera were grainy, and it was hard to determine what the piece of foam insulation had hit. The next day, NASA was able to analyze different camera angles and ascertained the foam tile had hit and put a ten-inch hole in the wing's leading edge. However, they could not tell the depth of the hole and let the mission continue as scheduled. That day Houston flight control informed Commander Husband of the issue. He was told to proceed with the experiments if there weren't any onboard system malfunctions and that that the ground staff would gather a team to analyze what could be done if anything. At that time, no one aboard *Columbia* was qualified to do a spacewalk to inspect the damage. The crew was told to monitor any temperature changes and continue as scheduled.

On the ground at NASA there was some concern about the hole after the pictures were examined, but unable to determine its size, NASA engineers produced several conclusions. None was better than another; it was just speculation as to what might happen. The ground crew even did some mock-up testing just to see what impact a piece of tile hitting the leading edge could potentially do. Using the shuttle *Enterprise*, they rigged a high-speed cannon to launch a comparable size piece of insulation from the oxygen tanks at five hundred miles an hour, striking the leading edge. They didn't like the results. On the tests, the same hole that appeared on the *Columbia* also appeared on the leading edge of the *Enterprise*. But what could be done? The crew did not have the expertise to do spacewalk (EVA) or make any repairs in orbit.

NASA was in a dilemma. The idea of a rescue mission was brought up. The shuttle *Atlantis* was scheduled to launch within two weeks. Working three shifts, the ground crew could have it ready within a week, and up on the *Columbia* there were enough supplies, food, water, and oxygen to last more than a week.

If the *Atlantis* were launched to go rescue the *Columbia* crew, it would require seven astronauts to space walk over to the rescue vehicle. None of *Columbia's* crew was EVA trained, so this possible solution was dismissed. The only solution was to hope the hole without the missing heat shields would hold up long enough and that the surrounding wing shields would absorb the massive heat build-up during reentry. The commander was informed of this information.

The next sixteen days went as planned. All the scientific experiments were conducted. All onboard systems stayed green, giving no indication of a problem. The crew was enjoying their stay, sending pictures down of them tumbling around in the weightless environment of space. Finally, after 250 times circling the earth, it was time to come home. The cargo-bay doors were closed and locked, and everyone donned their suits and prepared for the deorbit and reentry. Halfway around the world, the *Columbia* was put in the proper orientation to leave space.

With the computers in control, the shuttle began it deorbit burn to slow down and basically drop out of orbit. The deorbit was normal, but the reentry was the major concern. As we discussed in chapter 10, the *Columbia*'s nose required being raised to 40 degrees, which put it in position for the reentry. A 40 degree up orientation, exposes the bottom of the fuselage to the massive heat buildup as the craft descends, and the wing's leading edges take the brunt of the heat. Both the crew and ground monitors were watching many of the critical systems needed to go through reentry intently.

The crew reported a gradual increase in temperature, which is normal for all shuttles descending from space. Houston flight ground directors noted the same alerts. At around 250,000 feet, both the *Columbia* crew and ground crews noticed a spike in the temperature sensors, which is not normal. The crew reported the inside crew compartment was heating up rapidly. Commander Husband radioed saying the crew was concerned. At around 8:30 a.m. the hydraulic system red-lined, which is the maximum reading. As the shuttle entered the blackout period, there were no communications between the shuttle and the ground

Then around 8:58 a.m. there was another major alarm; the hydraulic system peaked. This is a critical system that controls the shuttle's airplane aerodynamics once in the earth's atmosphere—like the wing's ailerons, flaps, rudder, and anything that needed to be maneuvered and steered. At 8:58, the crew saw several systems were red-lined, indicating major system malfunction. Low enough for the ground, high intensity cameras could see the orbiter going across the sky.

On February 1, 2003, *Columbia*'s wing was so severely damaged from the increased heat built-up and pressure that it broke off and disintegrated. At approximately 9:00 a.m., two hundred thousand feet over Texas, the *Columbia* mission was over. The ground camera saw two major sections of the craft streaming across South Texas. The ground crew now knew the fate of *Columbia*. Debris was spread across Texas and Louisiana as it crashed. The other leading-edge heat shields next to the missing shields could not absorb the massive heat build-up, which caused the break-up and loss of control. All seven crew members perished.

After three years of training, many delays, and sixteen days in space, less than one hour from orbit it was all over. What a devastating loss. No one anticipated a ten-inch piece of foam insulation would be able to cause that much damage. But in space you cannot take anything for granted. In the weeks and months following the crash, somebody had to be blamed, and finally the NASA director stepped down from her position. Looking back many say the rescue effort should have at least been attempted. As the years have gone by, it is still something most can't agree on.

CHAPTER 12

Hubble Space Telescope (HST)

The desire to know what's out there is not new. For centuries, modern people have been glaring out at the night sky observing our solar system, trying to figure it all out. Ancient astronomers made all sorts of calculations, but no one was exactly accurate. Galileo was given credit as the leading astronomer of his time in the sixteenth century. Galileo proved certain elements about the makeup of our galaxy, but until the telescope was invented, we could only speculate what we saw looking at the night sky. And depending upon your location, your view may be hazy at best. Once the telescope was invented, we could only see so far. We could see thousands of stars, found out there were seven planets, the sun, our moon, the Milky Way, and other galaxies, but that was about it. We were extremely limited in our observation of the sky and beyond.

As mentioned before, humanity's quest for knowledge never stops. Only in the last century or so have we been able to build telescopes able to detect other galaxies and beyond, but the images have always been distorted and unable to distinguish fine aspects due to clouds, dust, and gasses on Earth.

Launching the HST

Only since the launching of satellites that have traveled or are traveling to other planets and solar systems, have we been able to detect certain things, but these satellites are traveling at enormous speeds and are not stationary, which still causes issues with sharpness, imagery, and focus. We built some large ground-based telescopes in rural areas—like the ones in Hawaii and Puerto Rico—to view the heavens with powerful lenses, but even those were obscured. A couple of telescopes were launched on top of rockets and from the shuttle to alleviate those issues with obscurity, but those still never really came close to what was expected. We needed something much larger and more powerful than we had ever had before.

Once again, our human ingenuity went to work. The Hubble Telescope (HST) was the answer. It took years to develop, build, and launch. Finally, after years of headaches, delays, and issues, it was launched on April 24, 1990, aboard Discovery STS-31. Hubble was a godsend. It expanded our horizons of knowledge. It allowed us to connect not only a few galaxies nearby, as we'd hoped, but galaxies thousands of light years away. Hubble now gave us the capability to see far away solar systems, their composition, and how similar they may be to Earth. Some say Hubble allowed us to view the beginnings of our universe.

Hubble Space Telescope floating away after being launched

As I have alluded to several times, anything manufactured has flaws and is not failsafe. Hubble fit that category as well. After all those years in development, many shuttle delays, and setbacks, there were lofty expectations.

I had been teaching shuttle operations for about three years in Sunnyvale, California, at the Lockheed plant where the Hubble was built. Several times I had gone to the manufacturing building to see the progress. Unable to get close because of the clean-room environment, I could view the work from a distance. Taking years to build, it was finally ready, but as always, NASA delays arose. And with the *Challenger* accident in 1986 and the subsequent issues with Morten-Thiokol concerning the SRBs, everything was pushed back.

Two years went by, and the Hubble sat in the plant, being maintained daily until NASA announced a launch day in January of 1990. Making its way down to KSC, it was ready for the launch under overly cautious conditions. I had hoped to see the launch but could not make the trip. The seven-person *Discovery* crew were anxious to get going, having prepared for as long as five years. Delay after delay pushed back the date for months with software issues, weather delays, and others. I could only imagine the frustration of the crew after all these attempts to launch. It was another thunderous ovation from the huge crowd who had assembled for the launch. The flight went off without a hitch, and within two hours, the cargo bay doors opened, and everything was in place to launch the Hubble Space Telescope.

The *Discovery* commander got the word to start the launch. Three mission specialists who had been training for years donned their suits, entered the chamber, and then went out into the hostile space environment. Since it was a large payload, the Canadian arm was used to assist in placing the satellite in orbit. The arm, attached to the payload, gently lifted the Hubble up and out of the cargo bay, where it seemed to be floating. As I mentioned earlier, with all satellites launched a timeline comes into play, and this was the same with the HST. So, time was of the essence. Still attached to the arm, the first solar array unfolded entirely. The crew, along with those on the ground, was ecstatic. The signal was given to start the second solar array deployment.

Nothing ever goes as planned, especially in space. As the button was pushed to start the deployment sequence, the array unfolded but stopped. Resetting the button did nothing. What had happened? Was it something in the software? Both the *Discovery* crew and the ground immediately tackled the issue. Hours elapsed with no answers. A partially deployed telescope, still attached to the arm while the shuttle traveled at 19,000 miles per hour could have led to a major disaster—not only for the Hubble, but the shuttle and crew as well. Something had to be done and soon.

With one array deployed and another partially deployed, the Hubble could not be reinserted back into the cargo bay because it was just too big. On the ground the software engineers were scrambling to produce a solution. Several fixes were attempted, but nothing worked. An engineer in Huntsville, Alabama, thought it might not be the software but a tension problem. The arrays unfold using a spring tension, which essentially pulls out the array to a fully deployed position. Bringing this to the experts, the engineer was not taken seriously at first since the first array deployed without issues. But with no other fixes, they set it up, and to their delight it unfolded to its correct position. Everyone was elated.

With the signal given, the Canadian mechanical remote arm was released. Now the Hubble Space Telescope after its inception, development, cost overruns, and years of delays was finally a fully functioning satellite. It was a glorious day. The next day, the crew prepared for deorbit. The Hubble was so bright that it was visible for hours. One last check by the crew found all satellite systems operating normally. The solar arrays, now fully deployed, kept the batteries charged. The Houston control center gave the word to come home. After one last pass around the globe, the deorbit burns and reentry went off without a hitch. Within two hours, *Discovery* was back at home. Everyone called it mission accomplished, and NASA had one of its major milestones. But don't count your chickens until the eggs hatch.

So many had gathered at Goddard Space Flight Center on that historic day, but when the first images came down in May, no one could believe what they were seeing. The pictures were expected to be sharp and crisp, but instead they were distorted and fuzzy at best. How could this be? What went wrong? It was initially thought the lens mirror was not curved to merge the incoming light. Calling it a million-dollar mistake by the press, many of the late-night TV programs were making jokes about it.

The overall project manager was devastated and dumfounded as to what to do. He had spent a lot of his adult life on this project, and it was a big failure. Many of the world's leading astronomers had been counting on the Hubble.

The manager gathered his team together to produce some workable solutions and fixes to save this expensive now piece of junk. His team worked for weeks on the problem along with Perkins-Elmer, another contractor, who developed the large lens mirror for the telescope and produced what they thought was the fix. Maybe the mirror was not polished to the right specs, which probably caused fuzziness. Polishing the onboard mirror created a major problem. The large telescope barely fit into the cargo bay when launched and was already in orbit. Launching another shuttle to go up and retrieve it, put it back in the cargo bay, and bring it home was not feasible. It was never expected the Hubble would need to be brought back to Earth for repairs. It was initially built to last approximately thirty-one years, only being serviced every three years in orbit.

However, returning the Hubble to Earth for a fix could have been attempted, but many felt this might do more harm than good. Contamination and vibrations during reentry might damage it beyond repair. Plus, the overall expense was astronomical, and who would foot the bill? So, this idea was dismissed. All around the country, NASA engineers and technical folks were hard at work trying to find anything that could work. Months went by with no viable solutions. Perkins-Elmer engineers (PE) were not convinced the lens was the issue. Their testing revealed that the light source reflecting off the mirror could be the cause

At the Huntsville Space Flight Center in Alabama, NASA and Perkins-Elmer engineers looked at the light entering the Hubble and found the light source did not match equally at the center as required. Further tests showed one of the focal points was off by a millisecond. With more tests, PE engineers were able to align the focal points where the light met all at once, and to their surprise it worked. But now it had to be done in space. The plan was to exchange the mirror to pinpoint the light for clarity. This required changing the mirror in orbit, which was a major undertaking that no one thought was possible. Kodak developed a second mirror as a replacement. But installing a second mirror, if needed, could only be done at the manufacturing plant and in a clean-room environment, not in space. There had to be another way. The NASA team and other engineers from Huntsville knew they had to solve the problem.

The Huntsville engineers put their best foot forward and created an instrument called Co-Star. Instead of replacing the mirror, a replacement box housing the four focal points used the Co-Star instrument to align the four points. The new box could be changed out in orbit during a spacewalk. The Hubble was built to exchange packages during servicing.

Now that a fix was eminent, still none of the NASA mission specialists had any experience in this area or were certified to do maintenance at this point. An immediate training program was established. This big of a repair job in space had never been attempted. Training for the space walks and repairs in orbit required new tools to be developed. The telescope, now in orbit, could not survive without maintenance, and it being useless was costing NASA a fortune.

With the entire world watching, the pressure to get it fixed was increasing. Spending millions of dollars on the HST project, there initially wasn't enough budgeted to continue the program. NASA asked the government for an increase in the budget, but with the tension in the Middle East and the Gulf War looming, Congress did what they always do and dragged their feet.

With enough public outcry, press coverage, and visibility from everywhere, funds from various sources poured in from across the country, private enterprises, academia, and others. Now with the fix and money available, the only hold up was getting it done. Once again NASA, being a government entity, red-tape and delays got in the way. Finally, after three years, the shuttle *Atlantis* was launched with seven astronauts and several space walks scheduled for the fix. The mission specialists were fully trained and prepared to do the in-orbit repairs.

Without the need for a mirror change, the Co-Star computer program and the box containing the four focal points would hopefully be the billion-dollar fix to save the whole program. The reputation of NASA, along with its major contractors Lockheed-Martin, Perkins-Elmer, and Rockwell, were riding on this and needed it to work. There were no second chances. Some on the ground called the replacement change out an easy fix, but *easy* is a word that should be deleted from NASA's vocabulary. It's just the opposite—everything is hard in space.

It was on! A shuttle was scheduled to launch in April 1993 to make the repair. Again, the entire world was watching. The EVA started out without any issues, and everything was on schedule. Three Astronauts participated in the fix. Using the Canadian arm for support, the first crewmember floated over to the HST, found the compartment where the Co-star package was to be inserted. Using special designed equipment unlocked the compartment hatch, notifying everyone the new updated package was ready to be inserted. The other two Astronauts, slightly push the package across and lined it up with the opening. A soft push on the Co-Star box, and she slowly went in, but only halfway. After two more attempts to get it in, they were hesitant to put any additional pressure on the box, fearing it could be damaged. What is Wrong? Many thoughts came to mind. Did the manufactures build it too large to fit the compartment? Were the pecs during initial construction were off somewhat. A lot of speculation crossed everyone' mind.

We did not come this far to fail. Someone produced the idea to wipe both the HST compartment and the Co-star box down, Maybe after several month now in space, the compart could have warped. No one knows. Another hour went by, and once the compartment and the box was wiped, the thumbs up from the Shuttle commander to start the process again. Using a micrometer as a guide, the Astronaut lined up the box, gave it a push, this time it worked. The box went into everyone's surprise. Everyone felt relieved. The hatch was locked, back on the shuttle and on the ground, no red lights, all in the green. Everyone was elated. The Canadian Arm released the HST, and she gently floated away and was a functioning satellite again.

Once back on earth the shuttle crew had did its part. Now it was only time, and within months all the HST major players were there to see the results. A loud cheer erupted from the room, letting everybody in the building know, the HST is no longer space junk, but a fully functioning satellite. Since becoming fully operational HST pictures changed a lot of things we thought we knew.

CHAPTER 13

ISS

What is the ISS, and what is its purpose? The ISS is the International Space Station. It is a floating laboratory for experiments that can only be conducted in space. It is owned by several countries, including the US, Germany, Canada, Japan, Great Britain, the European Space Agency (ESA), and Russia. Originally built in 1984, it was initially called *Space Station Freedom*. It went through many iterations, but due to funding and budget cuts, it was never built. Enough interest kept the program alive, and the name changed to the International Space Station. Launched in 1998 after several delays, the original cost was $60 billion, estimated to be $150 billion in today's dollars. It is the size of a football field with the solar arrays deployed. Weighing in at eighteen thousand pounds, it is a massive structure that was built and constructed entirely in space over a ten-year period.

Now owned and operated by sixteen partners, it is a highly praised asset and should last a total of thirty-one years with maintenance every five years. Still flying today, it is visible in your area periodically. It takes approximately ninety-two minutes to circle the globe and moves at 28,000 miles per hour in a low Earth orbit. Generally, there is a seven-person astronaut crew onboard, and residents stay for up to six months. With the STS program canceled, the only way to space now is aboard the Russian rockets and recently commercial rockets. To get to space, a three-person crew is taken up atop the Soyuz Rocket, launched out of Kazakhstan Space Center, near Russia.

The ISS is comprised of seventeen modules, or hubs, which are used for scientific studies, both exposed and unexposed to the environment of space. Other modules are used as living quarters, different countries' specific studies, docking facilities, storage, and other sections. On the outside of the structure, a Canadian remote robotic arm is attached to the ISS and used for various extra-vehicle activity (EVA or space walks). Most of the larger modules were brought up to space by the STS program during the last ten years before the program was canceled. The smaller sections were launched aboard the Russian Soyuz rocket, and all sections were assembled in orbit.

With the sections coming up periodically, numerous shuttle crew EVAs were used to assemble the main sections like a giant Lego set. Once the major pieces were assembled, then additional smaller sections were attached by the Russians, Canadians, and other countries. It finally became operational in 1999 and has been conducting scientific experiments ever since. Some of the experiments have been revolutionary and have improved life on Earth.

The ISS is organized in several sections. Covered in blankets to protect the ISS in orbit, the first part, which holds the ISS together, is the integrated trust system which is the backbone of the satellite/ISS. It holds the solar arrays, radiator panels to reduce heat, and other scientific experiment modules. The lower sections house the pressurized models where the crew lives and works. On the other end are the Russian modules where the cosmonauts live. On the opposite end is the US module, named Destiny; the Japanese module, named Kebo; and another section named Columbia for three countries: Canada, Japan, and the ESA. The various pieces were built by several countries and launched aboard the STS before 2011 then solely aboard a Russian rocket. The Russians were first to start assembling the Lego set.

The second major part was American, called Unity. It is a large module with six docking ports. Vesta is the Russian module with three docking ports. The Z-1 trust holds the equipment section and contained the pressurized mating adapter (PMA), #1, and #3. The solar arrays and radiator panels are attached to this section. At this point, there is enough functionality that humans can live and work in the modules. From November of 2000 there have been either astronauts, cosmonauts, or both living on the ISS. Over the years various sections have been rearranged to make it more comfortable. In 2001, additional spare parts for the station were attached to the ISS.

Seeing the need for more reach and maneuvering capability, Canada contributed with a large arm, attached to the side, to aid in EVAs and grabbing and maneuvering various parts. Then the US installed a Quest joint air lock. And not to be out done, the Russians installed their own air lock, called Pirs, on their end. Several more pieces were brought up and attached to the trust, labeled *S1* for Starboard, and *P1* for Port. Starboard and Port are old navy terms used to tell which part of a ship you are on.

International Space Station - ISS

Over the next few years, many other parts came up to complete the ISS. Eight large solar arrays were eventually installed to give the

ISS enough electrical power to keep it running. To extend the range of the Canadian arm and to maneuver parts and equipment to either side of the satellite, Canada built an extension to their initial arm. It is called Dexter. It attaches to the Canadian arm, giving it more flexibility. Other modules from the European Space Agency (ESA) and Japan were brought up last, which completed the build. These modules can do both inside and outside experiments.

Now that the ISS is fully operational, countries are requesting time aboard the ISS. With commercial launches like SpaceX becoming routine, crew change-out may be reduced to three months to give more countries their time doing studies in space. With the lifespan slowly dwindling down and close to its end, many countries are scrambling to get time on the ISS for their research.

Another vital piece, and very essential to the build-up, is the automated transport vehicle (ATV). It is like a small boxcar. It is about half the size of a container freight van, seen stacked aboard cargo ships. Fully stocked, it needs a large Russian rocket to launch into space. The automated transfer vehicle carries equipment and supplies, like food, medical supplies, or anything to support survival for several months in orbit. Carrying a maximum of twenty tons up prior to the STS program being cancelled, it docks to the American module of the ISS until the return trip back home.

Designed for a one-way trip up, the ATV's return trip down carries the trash and waste generated in orbit and burns up during deorbit. Stocked with food and medical supplies going up, it provides the nourishment required in orbit. Once in orbit, the ATV's solar array provides the power until it docks with the ISS. After emptying the ATV, they use it to store trash and unnecessary items.

Without a medical doctor on board, although some of the crew may have been licensed physicians in the past, keeping healthy is vital in orbit. To prevent bone density loss or osteoporosis caused by the weightless environment (a problem every crew member faces), an exercise bike, treadmill, and tension weights are installed. If not properly monitored, bone shrinkage will occur; therefore, a daily exercise session is mandatory. If emergencies arise, there is only so much that can be done. Launching an emergency vehicle is not something done hastily. It may take weeks to prepare a vehicle for launch, which is probably too late for any crew member needing medical attention.

Usually during elementary school presentations given by astronaut's, kids ask the crew members questions that adults are embarrassed to ask. One is: How do you eat in space? Other questions include: How do you sleep, and how do you go to the bathroom in space?

Eating in space can be difficult. Most of the food is prepared on Earth and, like eating military rations, they're called meals ready to eat or MREs. The food is usually dehydrated. There are no microwaves, toasters, or air fryers on the ISS. Adding hot water to the foods, like noodles, mashed potatoes, and spaghetti, softens the dish, and then the food can be sucked-in, chewed, and swallowed. The MREs are not especially tasty but are nutritious and provide all the vitamins and minerals needed. Certain foods, like toast and sandwiches, are prohibited due to crumbs clogging up crevices of equipment. Everything floats in space, so you need to be ready to eat once the dish is ready. Condiments like salt and pepper or ketchup and mustard are available, but they are not granulated. Liquids float around in a bubble and must be captured, or they will gently float away.

Sleeping in space takes getting used to. A sleeping bag like one used on a camping trip is provided for each crew member. Phone-booth type closets are used for sleeping and relaxing and are where the astronaut goes for privacy. In space you can sleep in any position—even standing straight up if you want. Trying to mimic sleeping on Earth, the astronauts say it takes a while to get used to. No gravity in orbit wreaks havoc, so trying to find a comfortable sleeping position is something learned very quickly in space.

Brushing your teeth is like back home; however, the toothpaste may float away if not properly attached to the brush. And after gurgling or rinsing, you have two choices. Either spit it out or swallow it. There are no showers on the ISS. Water is manufactured in space and considered a premium, so hygienic throw-away towels are provided.

Eliminating body waste can be an issue in space. A suction cup to capture urine from males, and a suction commode for solids is there for both. Accuracy is needed for solid waste due to the size of the commode, and if you miss there are wipes and other sanitary cloths to clean up any mess. All the waste is stored in tanks aboard the docked ATV, and on the return trip back home, the ATV burns up during reentry. This is a delicate element the crews soon master. Cleaning up waste once or twice is not a desirable task, so you quickly learn Robin Hood's skill of hitting the bull's-eye.

With scientific experiments being conducted daily, the crews are constantly doing some sort of activity. Restraint harnesses and a locked down foot-clamp mechanism keep them stationary in front of the equipment and keep them from floating away. In space you don't walk while in orbit; you just float to wherever you want to go. It is weird looking at the crew just floating about. A slight shove can propel one to his or her next destination.

Overall, the ISS has accomplished many things impossible to do on Earth. The ISS and its program have provided a platform to conduct scientific research with data, cooling, and crew available to support the research. Small, uncovered spacecraft can also provide platforms for experiments, especially those in a weightless environment with exposure to space. The ISS offers a long-term environment where studies can be done, some say for years, along with availability to scientists worldwide. Several types of research and experiments are conducted, including material, physical science, astrology, weather, human research, and many others.

Constant research is being done on the effects of long-term exposure to space if we are serious about traveling to other planetary systems. With planets and other solar systems light-years away, it would take months and years to reach the moon, Mars, Venus and beyond. Going further would take years. Can our bodies hold up to this long-term exposure to a weightless, no-gravity environment? Getting there quickly is the key. But present technology has not developed the warp speed needed to reach faraway destinations.

Like any other manufactured machine, the ISS only has a life span of thirty years, with a five-year maintenance schedule. Since it's been in orbit for twenty-one years, we only have ten years left. We are in a race to do as much research as possible before it reaches its end-of-life cycle. With so many countries involved with the ISS program and research, I am sure a bigger, better space station will be built.

Recently NASA and its contractors launched another significant milestone. The Webb telescope, which was under construction for years and designated to eventually replace the Hubble, finally launched on December 25, 2021. It is the largest and the most powerful telescope ever launched. In the next chapter, I'll discuss the Briggs telescope and another satellite that—although much smaller than those recently mentioned—is just as important.

CHAPTER 14

James Webb Telescope and TDRSS

As humans we want to know the ultimate answer to the question: How did we get here? The brand-new James Webb Telescope answers that question. It is much bigger and more powerful and can see light years further than our current Hubble Space Telescope. The James Webb Space Telescope (JWST) has been in development for twenty years as the replacement for the Hubble, and after years of development, it was finally complete. Once again it had to go through the NASA runarounds with setbacks, budget cuts, cost overruns, and delays. It was finally launched on Christmas Day, 2021.

The ten-billion-dollar satellite was launched from French Guyana at close to zero degrees latitude, near the equator. At or near zero degrees, more weight can be carried aboard. So far, the James Webb telescope has performed brilliantly and can look at the Big Bang or the beginnings of the galaxy. With the Hubble, there are limitations on its capabilities, and the Webb telescope picks up where the Hubble leaves off. Light traveling from galaxies and solar systems are now visible with this new telescope, giving scientists further knowledge of the make-up of our galaxy.

Using a variety of new coverages for both hot- and cold-type environments traveling to space, Beryllium was incorporated—a new material for the heat shields. Being a passive cooling system, the size of the shields are twenty-one meters long and fourteen meters across.

During construction, engineers worried about debris, ice, rocks, or anything hitting and puncturing the shields. Engineers also designed and sewed in rip stops every few feet to keep the shields intact. The heat shields can now allow the JWST to fly and orbit close to the sun where engineers can do research and conduct studies that had never been possible before.

At the John Hopkins Mission Operations Center (MOC) the engineers, along with engineers at NASA Goddard Space Flight Center, verify they command signals sent up to the JWST. These signals unlock the latch release mechanism, called pins. These pins unlock the secondary mirror. With the primary mirror being concaved and the secondary mirror being convex, alignment is critical. The telemetry sent back is checked for alarms. Without alarms, the ground engineers proceed with the next set of steps to fully unfold the satellite. After being fully deployed, the satellite travels to L1, its first parking orbit a million miles from Earth. Once there, an assortment of studies will be conducted before moving on to L2 and eventually the sun.

The five-layer heat shield protects the mirrors that always point away from the sun. With the mirrors pointing toward the cold side and the vastness of space, where temperatures can reach minus seven hundred degrees. These sub-minus temperatures create a huge engineering problem. Eventually the problem was fixed. The mirrors used are three times bigger than those on the Hubble, but they're also lighter. Made with eighteen hexagonal segments, each one is 4.3 feet in diameters. They were made this way so they could be folded up to fit into the housing atop the Arian rocket for launch. Using four camera packages, various spectrums of light are captured and analyzed. A new cryogenic refrigerator system had to be developed to cool down one of the cameras so it wouldn't detect its own heat source, which was another engineering marvel.

Not wanting to absorb any light from the sun, Earth, and moon, the JWST must be one million kilometers away. At that distance, there is no gravitational pull, which allows the satellite to remain in a fixed position. With its unique capabilities, what is next for the JWST? It's so accurate that an object the size of a deck of cards can be seen as far away as the moon. Detecting water is something else the satellite can do. And we may be able to solve that critical question of whether there is intelligent life out there somewhere. There is a lot of hope, and within the next year maybe some of these questions will get answered.

Although it's an engineering marvel, the JWST won't last forever. It's built to last for at least ten years, but there is no servicing or maintenance scheduled due to the distance away from Earth. With enough fuel to power the small rocket jets for approximately ten years, the satellite will eventually die. It will remain in its last orbit until it decays and becomes space junk or gravitation pulls it in and destroys it.

The James Webb Space Telescope has four key goals: (1) to search for light from the first stars and galaxies that formed in the universe after the Big Bang, (2) to study galaxy formation and evolution, (3) to understand star formation and planet formation, and (4) to study planetary systems and the origins of life. Using its large infrared mirrors, the JWST has capabilities never seen before.

I must also mention one of the most important but smallest satellites we have ever launched. Without these types of satellites, we could not get data back and forth. The tracking and data relay satellite system (TDRS) is so important, and without it, communication would be impossible. TDRS was the main work we provided at the Blue Cube at Onizuka Air Base near Sunnyvale, California, in the eighties and nineties before the program was eventually transferred to Colorado Springs, New Mexico, and elsewhere.

There is not one satellite but a series of them linked together to pass data using a ground-link system in the higher range of data, mostly used by the military. To make the system work, NASA created the spacecraft tracking and data acquisition network (STADAN). Three ground complex segments make up the system. These three are spread out with two in the US and the other midway in the Pacific Ocean. Large antennas, covered by a dome-shaped structure, located in southern New Mexico and Guam, called the remote ground terminal, and the naval computer and telecommunication complex at the GoDaddy Space Flight Center in Greenbelt, Maryland.

The TDRS system has been around for thirty years and is still going strong. It was initially launched by the space shuttles on several missions and then on unmanned rockets. As of two years ago, there are ten TDRS satellites surrounding the globe. More are planned to be launched to replace those that are reaching the end of their lifespans. Being a lightweight satellite compared to many others, several launch sites can be used.

The TDRS system is very essential to the communication of our satellites, ISS, and telescopes now in orbit. As more types of satellites and telescopes come on board and continue our surveillance, TDRS will take on a vital role in the upcoming years. The James Webb Space Telescope will open things we have never seen before, giving scientists answers to a multitude of questions and theories. With the ISS in its last decade of usefulness, there are talks to replace it, but it will take the combined effort of many countries—not just the US—along with funds probably in the millions and billions of dollars to build and replace the ISS, Hubble, and the JWST over the next twenty years.

CHAPTER 15

The End

I'll end by going back to the beginning. This book is about NASA and their little secrets that most people don't know about—telescopes and satellites, the various kinds, their functions, how they were launched, and the space shuttle program. The space transportation system, or STS, was created so humans could live and work in space. For close to thirty years, our satellites have been performing a critical function that provided not only atmospheric communication, but long-distance observation, to include surveillance during the Cold War era and still today. Sometimes the only way to orbit for our heavier satellites was using NASA shuttles.

The International Space Station, or ISS, has opened more doors for humans than ever before. In the last twenty years, major health improvements have come from research done aboard the ISS. These and other improvements have made our lives a lot easier. Hopefully with regular maintenance and upkeep, it might keep the ISS going past its expected lifespan.

I began chapters 1 and 2 with the day I'll never forget. At that time, I was working at the training center in Sunnyvale, California, providing training to USAF officers in charge of the satellite program to both Lockheed and NASA engineers and technicians and to the astronauts themselves. The space shuttle and the program had become routine once its first flight in 1981, so no one expected a major accident would take place. The first space shuttle built, named the *Enterprise*, never went to space, but five of them subsequently did all the heavy lifting. After many problems and delays, the STS program did that for which it was built.

With all the turmoil with the Soviet Union (now Russia) and problems in the Middle East due to oil and power, more communication- and surveillance-type satellites were urgently needed. It was vital for shuttles to launch and provide these assets in orbit. The *Challenger* accident, which never should have happened in the first place, set us back. With almost fifty states participating in some aspect of the shuttle, the crash affected the whole country.

In chapter 3, I discussed the aftermath of the *Challenger* accident. With all the shuttles grounded, NASA was scrambling trying to find out what caused the crash. Inquiries, inspections, and more delays took years. In the meantime, things were escalating in the Middle East and beyond. A backlog launching satellites was common. With only two launch sites, one at the Kennedy Space Center and the other at Vandenberg AFB, launch schedules became hectic. Russia saw our need to launch and offered their sites. We accepted their offer to launch payloads but did not launch our surveillance satellites from their pad, only using KSC and Vandenberg for DOD and classified launches.

NASA's workload suffered with layoffs. Being in the training center without the shuttle flying, our jobs were in jeopardy, so we started developing unmanned-launch training programs. It kept us going until finally after so many delays the shuttles were back up and running. Taking on the unmanned-launch vehicle training and now with the shuttle back flying, our productivity picked up twofold.

I mentioned the folks at the training center and the Blue Cube, who were responsible for day-to-day operations and maintenance of many of our satellites. We then discussed the shuttle and its capacities and how it became a workhorse for NASA. I discussed how we transported the shuttle on Earth from one place to another using the modified 747. We talked about the astronaut training they must go through, which could take years. We discussed EVAs and what they entail. I mentioned the lifecycle the shuttle goes through with its cradle-to-grave concept—how it is readied at KSC in the VAB, moved to the launch site, launched, and after performing its mission, deorbits and lands back on Earth.

In chapter 11, I talked about the *Columbia* disaster, which set NASA back a few years, throwing everything in turmoil again. In chapter 12, I talked about the Hubble Space Telescope—things like its mission, lifespan, cost, durability, fixes, and updates needed to keep it functioning.

Chapter 13 covered the International Space Station or ISS, which is another engineering marvel that's still flying as we speak. As I mentioned several times, anything man made has a limited lifespan. And while the ISS has already been going strong for twenty years, it will soon come to its end in another ten years. However, with some repairs and updates, we may be able to keep it going longer than the ten years projected.

Finally in chapter 14, I covered the new James Webb Space Telescope, JWST, and one of the most vital satellites, the tracking data relay satellite system used for communication. The new JWST, some say can do things we were never able to accomplish before when it comes to looking at the stars, solar systems, and beyond. Using infrared technology, we can see clearly past the dust and debris.

The tracking data relay satellite system is not one satellite, but ten or more, placed in strategic locations surrounding the globe, transferring, and sending data back to our ground stations. It is so vital to the security of our nation. Without satellite communication, it would be like the stone ages again. Instant information is vital for making high-level government decisions.

This book is not intended to badmouth NASA or any of its subsidiaries, but what I learned over the years I feel should be shared. With the canceling of the STS program in 2011, two major accidents along the way, and billions of dollars spent, many NASA employees have a sense of pride working for one of the most innovative organizations in the world. NASA still occupies a prominent place in space travel, despite civilian spacecraft companies making a leapfrog approach to dominate the space industry. The future of getting to space is still considered to be our last frontier, and everyone is stepping up to do its part.